Prateek Goorha · Jason Potts

Creativity and Innovation

A New Theory of Ideas

palgrave
macmillan

Prateek Goorha
Independent Researcher
Boston, MA, USA

Jason Potts
Royal Melbourne Institute of Technology
Melbourne, VIC, Australia

ISBN 978-3-030-06930-8 ISBN 978-3-319-94884-3 (eBook)
https://doi.org/10.1007/978-3-319-94884-3

Softcover re-print of the Hardcover 1st edition 2019

Cover Design by Tom Howey

This Palgrave Macmillan imprint is published by the registered company Springer Nature Switzerland AG
The registered company address is: Gewerbestrasse 11, 6330 Cham, Switzerland

Creativity and Innovation

Preface

It is safe to assume, since you are reading this book, that you share with us an interest in understanding how ideas work. There are a great many very good questions one could ask about ideas: Where do ideas *really* come from? Why is it that ideas seem to come differently to some people? What does the process of innovation *look* like? What are the features of an idea's *context* that actually matter? What, in actual fact, is creativity?

Perhaps an equally good question might be: There are, after all, many excellent books one can read on these topics, so why write another? Let us answer this last charge first.

While it is indeed true that one can read a variety of books on the topics that concerns us here, our objective was to write a book that proposes a simple and intuitive theory—a useful mental model—that one can keep in the back of their mind when thinking about ideas, regardless of what aspect of ideas interests them. We think there is immense value in having a common platform that we can all share in thinking about ideas, whether we are poets, business managers, physicists, artificial intelligence systems designers, or athletes.

If you think, much as both of us did at the outset of writing this book, that a theory of this immense ambition is entirely impossible, futile even, your misgivings would indeed not be ill-guided. How, after all, can anyone possibly propose a 'theory' about ideas—let alone *all* ideas—in the space of a relatively short book that remains accessible to *everyone* who has any interest in ideas?

We struggled mightily with this question, and especially about the generality and accessibility we sought for the overall theory, but decided in the end that the task was necessary enough to warrant attempting, particularly since we felt that our theory held the potential for achieving what we saw as its objective.

We live in a world of exponentially increasing ideas, and the stakes are simply too high to not have some basis for a common understanding for how ideas work.

Toward a Universal Law

Let us set the mark for the broad subject matter of this book by considering the message from a famous letter that Thomas Jefferson wrote in 1813 (Founders Online 2018).

On the topic of ideas, he began by conceding that they are, indeed '…the fugitive fermentation of an individual brain…'. He then made his position clear by averring that if '…nature has made any one thing less susceptible than all others of exclusive property, it is the action of the thinking power called an idea…'. However, once the idea is revealed publicly, Jefferson argued that the idea then '…forces itself into the possession of every one, and the receiver cannot dispossess himself of it'. It is in this letter that his often-quoted phrase lives. He explained that '(h)e who receives an idea from me, receives instruction himself without lessening mine; as he who lights his taper at mine, receives light without darkening me'. In the course of his life, he never sought a patent for his inventions, for it was his conviction, stated in this letter that '…ideas should freely spread from one to another over the globe, for the moral and mutual instruction of man, and improvement of his condition … like fire, expansible over all space, without lessening their density in any point…'.

The letter was occasioned by a fascinating issue: The prolific American inventor, Oliver Evans, had been granted an extension to his patent on the use of three technologies that were exceedingly useful for their labor-saving benefits at mills. However, before the extension could be granted, the technologies had been adopted into practice by several parties, much to the chagrin of the irate inventor, who now sought to be compensated. As someone instrumental to the patent office since its inception in 1790, Jefferson had been requested to provide his perspective. As remarkable as his assessment of the case was—in sum, he concluded that two of the three ideas were hardly even worthy of a patent in the first place—it is this quote from it that captures a great deal of the inspiration for why we undertook writing this book.

It is clear, even to someone who may know nothing of the man, that Jefferson had thought long and hard about ideas, and felt keenly that they held a special significance to humanity. The characteristics of ideas he enumerates inspire thought and are some of the key elements of our theory as well. Our theory, for instance, is in accord with Jefferson's observations that ideas ought to be seen, first, as 'fugitive fermentations of an individual brain'; that, by nature, ideas resist being the subject of 'exclusive property'; that the revelation of an idea 'forces itself' into general possession; and that ideas are 'expansible over all space'. These are precisely the sort of characteristics that one would have in taking an idea-theoretic view of the world. Jefferson's theory of ideas rather clearly seems to be that, while ideas may be developed by individuals, they resist being bound by the confines of an individual's mind. They possess the impulse to propagate into the awareness of a broader society and permeate in all directions.

The present represents the perfect time for all of us to be thinking just as deeply about the characteristics of ideas; we feel that the ability to think about ideas, a form of metacognition perhaps, makes us all better and more conscious thinkers. Having an idea-theoretic view of the world forces us to think beyond our usual concerns and specializations. It readies our mind to conceptualize the impact of ideas that have much wider implications than we might imagine. This is a considerable challenge. Jefferson, polymath though he was, could not have imagined how his perspective on ideas might one day apply to algorithmic and robotic artificial intelligence, though he would likely have strident and cautionary advice for us, even as he considered their value to the progress of humanity. The fact is that artificial intelligence, in its numerous guises, thrives on information at a pace that is simply beyond the grasp of humans, and so the point worth pondering is whether the force of logic that drives the value of ideas for human progress will now multiply the pace of progress for artificial intelligence to levels that are unfathomable, and perhaps even unintelligible and uncontrollable.

The disruptive abilities—in both the good and bad sense—of automated bots in finance, media, social networks, surveillance, marketing, and politics alone are a mere glimpse of what is possible in the near future in terms of how quickly artificial intelligence has already begun the journey of expanding its influence over ideas 'over all space' of human concerns. Some such disruptions are, indeed, attributable to insidious human intention, though others are also based on human ideas forming on the basis of erroneous inferences made by the AI technology. And yet, an unpleasant eventuality is not inevitable if we can first understand the interconnected and complex manner in which humans interact with their environments through all sorts of ideas and then design algorithms that are cognizant of the wider implications of AI, far from their intended purpose. We hope that our theory, by producing a broad enough framework for understanding the mechanisms for ideas, will be useful for the conversation in this regard.

A Bond Across Time

Ideas are a public good in a sense that is far deeper than is meant by the average economist: the mere fact that they can be shared without diminution. They are 'public' because they endure beyond us—indeed, beyond even any limited definition of a group of people that we can envisage when we consider what our own community entails. And, as such, ideas constitute a bond across the unfathomable stretches of time and between unseen minds. Our example in Part III of this book on 'Why We Play' endeavors to explore precisely this feature of ideas for the context of exploring the ideas that gave rise to the kinds of sports we play or follow—ideas that we take for granted.

While one may reasonably take the view that one of several characteristics are 'uniquely human', we do not believe that thinking about ideas—metacognition, as it is called—can be counted among them. Ideas are the elementary building blocks of thought, and so, to the extent that we grant that a variety of animals are capable of some form of improving upon their processes for thinking, we extend the purview of ideas beyond humans. Similarly, with modern machine learning algorithms, such as those being developed by deep learning theorists, ideas are now unambiguously not merely creatural inputs in the making of intelligent machines, but, increasingly, novel ideas are also the outputs that we see and expect from such applications.

This is a humbling thought. In a broad sense at least, ideas themselves can't be seen as unique to us, let alone be seen as bound by the ersatz organizational labels that we may place upon them in any given social, political, economic, or cultural context. This is hardly a new observation. Richard Dawkins made a similar observation more than four decades ago in his groundbreaking book, *The Selfish Gene*. Ideas, he suggested, endure beyond the limits imposed upon them by the individual. He remarked that '(w)hen you plant a fertile meme in my mind you literally parasitize my brain, turning it into a vehicle for the meme's propagation in just the way that a virus may parasitize the genetic mechanism of a host cell (Dawkins 1976)'. We will largely sidestep the word meme in this book; it has gathered several meanings ever since its introduction, not just on account of the field of memetics, but also through popular culture. Instead, our focus will simply be on ideas.

The endurance of ideas is a key theme in this book and of the theory that we outline in its pages. However, it is not just that ideas have remarkable endurance that makes them so interesting. It is also that they have a rather unique hierarchy which frames the manner in which they both endure and alter. This alteration is often seen as progress and often lamented as retardation because we impart the *course* of an idea's alteration significance, and we actively participate in engendering alterations to ideas and affecting this course in specific ways. This process of an idea's development is a second key theme in this book.

Ideas can be found in scientific knowledge, to be sure, but they can also be found in behaviors that are beyond the scientific domain, not just because they are often 'unscientific' but also because they may not be explicable. A Newtonian Law of Motion has as much elegance for its explanatory power as a Shakespearean sonnet has for its universally evocative allure, and yet both stand as ideas. Indeed, as we wish to suggest in what follows, they both stand as representations of *several* ideas, some of which are not just similar, but necessarily *identical*. We, therefore, propose a theory that does not favor any particular genre of ideas, but attempts to look at the broader mechanisms for all ideas generally.

This book's claim is that there is fundamental value in having a theory of ideas as a workhorse to guide our intuition on any idea, wherever it is found. Just as there is value in the formal theories that guide our understanding of scientific knowledge and in theories that inform our understanding of social norms that affect our behaviors and of those around us, there is value in having a theory that can guide our explorations of ideas, regardless of where they appear.

Without the rigor imparted by disciplinary scholars on the creation of principles that aid our understanding of phenomena, there would have been precious little basis for a book of this nature, not least because we will freely use concepts spanning a wide range of disciplines in developing our own theory of ideas. However, it is perhaps equally necessary to have some broader framework that permits us all the ability to step back and see the forest without forgetting that it is the careful understanding of trees that has enabled us to appreciate the forest all the more. A theory of ideas such as the one we present, therefore, is meant for the generalist who doffs her hat at the dogged specialist, but prefers to examine the whole message.

A Note on Reading This Book

First, this is not meant as an academic book for a professional audience. The overriding purpose we had in laying out a theory of ideas was to provide a simple and intuitive method for thinking about ideas for as varied an audience as we could imagine.

We have, therefore, attempted to present the argument as simply and clearly as we could manage, but also without any attempt to dilute the crux. That said, since we are trying to provide a robust framework with our theory, it was not entirely possible to smooth all the arguments by extracting all the terseness that accompanies proposing a model.

Chapter 1 presents the case for looking at ideas as the most basic unit of analysis, and we suggest that all readers begin there. It also presents a few applications of the theory in broad strokes. While we more expansively develop an application for our theory in Chapters 5 and 6, these briefer applications introduce some of the concepts of the theory and can be referred to again as the reader makes progress through Chapters 2–4 where the theory is detailed more fully.

Chapters 5 and 6 consider an application of the theory to finding the idea antecedents for sports. We concentrate on a class of sports, broadly defined as bat and ball games played in teams. The allure of looking at an idea from the lens of the theory we present is that it permits (indeed, it encourages it!) looking at how an idea's context affects its evolution. It draws attention to the process of how ideas might be motivated by some unrelated and broader spectrum of ideas, and how they might, in turn, become critical to other such seemingly unrelated ideas. Sports are, of course, a fantastic microcosm for a range of ideas and, so, our purpose in selecting this application is to explore some of this variety and its interconnections.

Finally, a word about the title. We realize that calling our efforts a *New Theory of Ideas* may sit somewhat uncomfortably with some, who may even perceive our phrasing as smacking of grandiose pretensions. To a 'real scientist' calling something a theory suggests that it is technically sound and scientific, and perhaps even that it is fairly definitive. To a philosopher, theories of ideas—Descartesean, Lockean, Humean, and so forth—are the subject of intense and unresolved debate.

At the risk of causing some discomfort to such individuals, we thought that the title actually befits our intentions rather well. While we do not claim to have made a sensational new scientific discovery here, we absolutely do wish to signal to the general reader that what follows has helped our own understanding of ideas and the process of innovation immensely. And we hope our readers find similar benefits from our work, as well. It is true that a theory should be seen as a set of tools that enable an individual to understand some phenomenon and to be able to use it to make some predictions. One of our favorite economists, Ronald Coase, once said that 'a theory is not like an airline or bus time-table. We are not interested simply in the accuracy of its predictions. A theory also serves as a base for thinking. It helps us to understand what is going on by enabling us to organize our thoughts (Coase 1982)'. Our theory of ideas is meant for precisely that ambition—a base for thinking about ideas.

Boston, USA Prateek Goorha
Melbourne, Australia Jason Potts

References

Coase, R. (1982). How Should Economist's Choose? In *Ideas, Their Origins and Their Consequences: Lectures to Commemorate the Life and Work of G. Warren Nutter.* Washington, DC: American Enterprise Institute for Public Policy Research.

Dawkins, R. (1976). *The Selfish Gene.* Cambridge: Oxford University Press.

Founders Online, National Archives. (2018). Thomas Jefferson to Isaac McPherson, August 13, 1813. Last modified on April 12, 2018. https://founders.archives.gov/documents/Jefferson/03-06-02-0322. Accessed on May 6, 2018.

Acknowledgements

I wish to acknowledge my co-author, Jason. Over the course of our association he has encouraged me to think in a manner that has been critical in the writing of this book. He has helped me develop the viewpoint that not just favors the evaluation of merit in ideas that force me well outside my favored modes of thinking and cherished explanations, but encourages pursuing such ideas as a habit.

Prateek Goorha

Contents

List of Figures

Part I

Introducing Ideas

1

The Most Basic Unit

As humans, we measure extensively. We measure deterministically and probabilistically. We seek measurements for our height and weight and the temperature (be it for indoors or outdoors, or merely as conversation starters) just as often as we desire to learn the odds of winning a gamble. We try to measure intelligence with GPAs, SATs, IQs, and a laundry list of other tests. We measure value with all manner of explicit and implicit prices. And we measure copiously and religiously in all things pertaining to sports.

We also measure hypothetically. So deep does our penchant for measuring things go that we like to create our own units of measurement where none exist or make any sense. We wonder how much our friends like us by how much they desire to engage with us; how ardently our significant others love us by how deeply they look into our eyes or how much thought they put into a gift; how committed we are to some cause by how much fervidness we exhibit, and even how our current behavior entitles us to receive benefits in the afterlife.

The reason we measure some idea is, of course, that we seek information that often helps us assess, compute, or predict some other idea. We rely on mental models that associate ideas in a desire to draw ever clearer links between a range of ideas and the information they can

P. Goorha and J. Potts, *Creativity and Innovation*,
https://doi.org/10.1007/978-3-319-94884-3_1

confer; this is a visceral need for humans *because* it is also an infinite process. The infinitely expansible character of ideas turns all ways of thinking into compromises from exactitude to varying degrees of certainty. Information on ideas can often have critical survival value, and so there appear to be sound evolutionary reasons for being able to improve our measures for the information inherent in ideas. It is no surprise then that a great variety of animals have some ability to measure as well. Some basic facility with counting, for example, is now thought to be a fairly widespread ability among so large a number of animals that a biological basis for it is suspected (Tennesen 2009). It has even been shown that some primates may have the ability to understand the basic principles of barter (Paulos 2011).

Yet, measuring in the pervasive manner that we do is, by and large, a uniquely human attribute.

The reason for this distinction is that humans, regardless of their particular pursuits and interests in life, are not merely satisfied in being able to apprehend ideas, but we are also interested in being able to *clarify* our ideas by extending them with other ideas, developing them into better ideas and merging them with others in creating new ideas. It is this process that drives us to look beyond the limits of our own minds and to to those of others in our community; as such, it is an essential basis upon which a variety of communities can hold significance to an individual.

In Yuval Noah Harari's remarkably broad and insightful book, *Sapiens*, one of the central arguments is the observation that humans have established themselves as the predominant species on the planet by dint of their capacity for fiction and ability for abstractions from reality (Harari 2015). He argues that these characteristics have crucial significance in enabling humans to cooperate in ever larger communities. The human ability to make abstractions in their ideas in useful ways is, indeed, remarkable. Daniel Dennett, in his influential book on the philosophy of the mind, *The Intentional Stance*, suggested that an understanding of the characteristics of a system and the behavior of people is open to subjective interpretation, but that these interpretations rely on the premise of shared beliefs and rational thinking. He defined three levels of abstraction that humans can take. The 'physical stance' would be applicable to phenomena that rely on laws that are useful in making reliable predictions. These largely comprise physical laws that permit consistent and reliable predictions. At one level of abstraction higher than that is the 'design stance', which takes the physical stance as a given and allows understanding objects that are built or designed on their basis; the physical laws are largely subsumed within the design. At the highest level of abstraction, Dennett proposed the 'intentional stance', where assumptions are made on the beliefs, desires, and the objectives of others (Dennett 1989).

We find ourselves in ready agreement with these observations, and with the aid of our theory will seek to articulate an intuitive and fundamental theoretical process for ideas; doing so can suggest why it is that some ideas can exist in isolation where others facilitate the creation of larger groups. What we gain by undertaking this exercise is a broader understanding of how ideas develop, often by blurring the lines across levels of abstraction and between fiction and fact while seeing both as waypoints along an unbroken process. Since this process is variegated and multidimensional, it is, we think, easy to perceive it as being marked by distinct 'phases' when seen with the help of a chronological framework.

It is perhaps useful at this juncture to also suggest how our view differs from the tradition of research on the Theory of Mind, or ToM, which examines how humans go about understanding the state of their own, as well as someone else's mind, including their desires, ambitions, objectives, and beliefs. While there are some obvious similarities in our goal to understand an individual's ideas, principally our intention is more modest; we wish to examine the process by which there is an innovation of ideas and, as a result, we emphasize the primacy of ideas over the state of an individual's mind, or how it develops and differs from its peers. With our theory, we are interested in studying the interconnectivity over time and space of all forms of ideas; the interesting topic of the significance of their states, as represented within the mind of any given individual or set of individuals, is of secondary importance for the purpose of our theory. There is, of course, a whole gamut of issues pertaining to interpersonal relationships and perceptions that are then diluted in our approach, but the benefit is that of underscoring the fact that ideas have an independent life and deserve examination in their own right, free from direct attachment to humans.

Withal, by understanding the nature of ideas we are able to use them as a key unit of analysis to examine both creativity and innovation. It is, therefore, unsurprising that the ability to measure ideas to yield usable information is an attribute that we have been at pains to engineer, codify, and intensify with the help of the machines we create. We tend naturally to think of the age of computers as being pivotal in terms of introducing measurement and computation to machines, but this underemphasizes the fact that self-regulation in machines has been a characteristic that was routinely seen even in the earliest of steam engines at the beginning of the eighteenth century. What is striking now is, of course, the extent to which machines have permeated their influence in our lives and the degree to which their capabilities exceed our own.

Evolution has indubitably provided humans a significant advantage over other species by increasing our relative skills for the processing of information over millennia. Taking the cue from evolutionary logic, humans are now at the verge of building adaptive evolutionary computation systems that can outdo both primitive machines and humans. And this evolution in machine learning is happening at a rate that far outstrips anything in the story of our own evolution.

1.1 Communicating Ideas

There are ostensibly between six and seven thousand languages in existence today. The discrepancy in counting up exactly how many there actually are is understandable for at least two very good reasons. First, many of the languages have no written tradition and are spoken by small populations, often just single-digit cohorts of native speakers who live in remote locations. When such languages will go extinct, they are likely to vanish without leaving much of a trace. Many do, despite the laudable efforts of intrepid linguists who take the time to visit the last remaining speakers of such languages and desperately compile records. Second, the rate at which languages are disappearing is quite staggering (Wilford 2007).

It has been estimated that three or four dozen languages are going extinct every year, their traditional speakers opting to learn a more predominantly used language and assimilating into a larger society (Pagel 2012). Naturally, that rate will not continue forever, and some languages will withstand the exodus of its speakers simply by being able to rely on maintaining a large enough community of speakers or on a significant volume of cultural resources that employ their usage. Nevertheless, it has been suggested that in another hundred years or so, the total number of languages remaining will likely decrease by a factor of ten. The languages that do survive will generally be relatively simpler than their predecessors (McWhorter 2015).

Reading about this alarming state of affairs makes us think of what needs languages suffice and why it is that at least some of those needs may have required more languages in the past than might be called for in the future. It appears that one can consider that question from a number of different perspectives, including the economics of how languages facilitate transactions, the sociology of how languages serve as instruments of cultural transmission, the psychology of how languages might serve as modes for cognition and for regulation of behavior, and so forth. One of Noam Chomsky's key contributions to linguistics was the fascinating proposition that humans are born with a universal grammar that is invariant of the language that they eventually speak. Steven Pinker refined this premise in favor of an evolutionary expedience for humans to be uniquely endowed with an instinct for language (Pinker 2007). While we see this field of enquiry to be a very useful step in the direction of the generality we seek in understanding ideas and innovation, the simpler observation that perhaps has the broadest intuitive appeal belongs to John Locke, who noted that words are nothing but the markers of the ideas that the speaker has in her mind.

When we think about most everyday objects, our ability to refer to them in context is based on their essential features. A hammer devoid of its recognizable features—perhaps one that has neither a handle nor a head—may, of course, still be used as a hammer. However, understanding this other type of 'hammer' would likely require a great deal more effort from a community of people that has already clarified what the essential features of a hammer are over the years, and now relies on this shared definition implicitly. The members of this community would now need to re-evaluate how the features—the 'component' ideas—that this newfangled hammer represents permits it to also function like the hammer that they are familiar with. They would, in all likelihood, find it convenient to think of the features of this new hammer more abstractly and then see how these abstract features enable it to leverage the *same* ideas—the same principles of physics and mechanics, the same ergonomics and feel—that their own trusty hammer represents. Such abstraction can, of course, vary in difficulty; indeed, if the new hammer looked like a feather suspended by a set of springs, we may even find ourselves at a loss to understand how any of its features might reasonably be translated to those of the trusty hammer in our tool chests.

Note that this example need not have anything to do with languages. It is, of course, true that when we consider the speakers of different languages, we add an interesting layer of complexity in the communication of ideas. However, it is useful to examine the role that ideas play on their own, free from the language that is used to express them.

Generally, the point is simple: Our perceptions of objects around us rely on 'labels' we ascribe to sets of ideas that represent those objects. All objects that are commonly known across some group of people have recognizable features, and while there may be some variance among them in the manner they perceive these features, there is broad consensus on their relevance to the object. This defined group of people understands the object's function and shares a language that expresses the ideas of its features with sufficient precision for the object's label alone to succinctly convey some acknowledged purpose and intent. Through their use of the object their understanding of the object's characteristics, whether good or bad, grows. When the object is then introduced to a second group of individuals who have previously never encountered it the label loses some if not all meaning; commensurate with the degree of ignorance of the object within the second group, the first group would then need to present the ideas that define the object so that it can be 'rediscovered' by the people in the second group. These individuals wonder what the purpose of the object might be, and perhaps even how it contrasts with objects that might be serving similar functions in their own lives. When there are at least some parallels, the labels can be adjusted in meaning. When there are no preexisting labels for the object, one may need to be introduced into their lexicon.

Indeed, this process extends to our interactions with machines, as well. The clearest example comes from the idea of supervised machine learning, which involves a group of individuals teaching a machine to learn the significance of a label assigned to a very large set of objects that all display the same features that the label is meant to capture. Once the machine learns this mapping, it essentially reverse-engineers the mental models that we have developed to aid our own understanding of which features we instinctively associate with the label. Naturally, the better the machine learns this mapping of ideas that certain features of an object represent with the idea that the broader label connotes, the better it is then able to use this information to identify other similar objects on its own in more complex environments.

A key observation we might make is that labels come and go, but the 'language' that any two groups—be they two sets of people, or people and machines—use for communicating with one another is based on the most basic unit of communication: *Ideas*.

However, ideas are hardly universal, regardless of what type of language is used to express them. When we think of the most universal of languages, mathematics is the one that usually comes first to mind. Yet its teachers frequently write about the challenges they overcome in making their curricula intelligible to new students from different backgrounds (Tevebaugh 1998). So, the universality of the 'language' of basic mathematics relies on an inherent translatability of the assumptions made on how its core foundational ideas are understood by one group of its students compared to another. Once that translation has been made, basic mathematics really does become a universal language. Labels in mathematics, then, carry over from one group to the next relatively easily because their component ideas usually translate with high fidelity.

The assumptions that are made to undergird an idea in mathematics outline how a set of core ideas relate to one another using a 'schema' that can be shared across a set of people (Axelrod 1973). Different groups of people may hold different assumptions even over largely the same core ideas, requiring a translation across the different schemas before they can be made to communicate with one another. As mathematics becomes more and more complex, the probability that the schemas between two different groups begin to diverge also becomes greater; the expanding number of ideas that the schemas represent makes the task of translation between them more and more involving.

1.2 Perceiving Ideas

It is no surprise, then, that the concept of some schema of linked ideas is integral to understanding how an idea is perceived at a fundamental level. It has been used across a wide variety of fields: in psychology, where it appears to have been developed, but also in sociology, information science, political science, innovation studies, and a few others besides. The reason for this popularity is that schemata help breed a sense of familiarity for objects across the group of people that share them. The manner in which an object is understood depends to a large degree on the way that a defined set of related ideas shapes an individual's cognition. This observation has a number of profound implications.

First, current perception of an idea, constructed as a schema of ideas, is almost never devoid of our past experiences (Bruner 1957). We are almost hopelessly destined to introducing the ideas that we hold as true or relevant about the world from past experience into the framework of our understanding of any new experience of an idea (DiMaggio 1997).

What this suggests about our most vivid and most cherished ideas is that they are valued by us as a direct result of their reliance on an expansive and deep structure of ideas that have served to burnish them into their current form. This point is illustrated quite thoughtfully in Fred Hoyle's rather brilliant science fiction novel, *The Black Cloud*. In it, the highly intelligent protagonist is given a chance to learn the secrets of the universe from an alien entity of infinitely greater wisdom by a sort of download of the information directly into his brain through the medium of vision. Sadly, the experience kills him, his brain unable to cope with the requirement of relearning what is actually true about the universe without first having the benefit of unlearning what he already knew as truth or fact; death by extreme cognitive dissonance, as it were! As he passes on, he poignantly remarks that the simpleton gardener in the story would have endured the experience far better than he had.

The second implication pertains to the fact that the degree to which any new object has features and characteristics that represent ideas that readily inhere to the features and characteristics of objects that we are familiar with will impact our willingness to accept and adopt the new object (Hargadon and Yellowlees 2001). No matter how progressive and savvy we may think of ourselves as a species, it is an undeniable fact that we prefer some degree of familiarity even within novelties.

In other words, we use ideas as hooks between the known and the unknown. The application of this insight to innovation is intrinsic: An object that embodies ideas that are not premised on ideas that are derived from some extant schema is less likely to be adopted than one that appeals to ideas from within that schema, even if it does attempt to modify them.

Psychologists, for example, have suggested that a two-stage model for visual perception characterizes early learning: young children tend to find the familiar in new information, and seek the novel only after the task of recognizing the familiar ceases to stimulate excitement (Wetherford and Cohen 1973). Generally, this observation forms a basis for much of product design and is, perhaps unsurprisingly, one of the most enduring principles in the study of product marketing (Rindfleisch and Inman 1998). An object's features need not be limited to visual characteristics; they include names, and, there too, such biases in perception based on familiarity have been shown to exist (Park and Lessig 1981; Rindova and Petkova 2007). Whether the name of the object sounds familiar or not influences whether it is perceived as being a safe bet or as unacceptably risky (Song and Schwarz 2009).

A different and perhaps more colorful analogy of this process comes from the idea of comparing ideas with nutrients and knowledge with living organisms. The essential insight is that, just as organisms do not directly adopt food from the environment without processing it to suit their particular needs, a body of knowledge also processes ideas that are derived from an environment (Weiss 1960). Rarely are these ideas directly stored as immediately relevant to the body of knowledge. More than likely, such ideas would first be subjected to analysis in order to determine their relevance to that body of knowledge before being absorbed by it and serve to augment the accumulated knowledge.

1.3 Key to Immortality

That neatly brings us to the key 'idea' (admittedly a word that we must use advisedly in this book, lest we cause an inadvertent confusion with the subject of the book itself!) that is the basis of this book:

Ideas seek to become immortal.

While this premise may perhaps sit uncomfortably with some for sounding somewhat mawkish or needlessly anthropomorphic, it is at the heart of the theory that we shall attempt to construct for ideas in this book. This simple assumption on the nature of ideas enables us to impart analytic power to our theory of ideas and permits us to develop a useful transdisciplinary focus in understanding our behaviors and our worldviews as individuals and as social groups.

Perhaps it would be more correct to say that ideas seek to endure in the face of competition from other ideas since the fact is that successful ideas *do* endure. They routinely have lives that exceed those of individuals, institutions, societies, and, sometimes, even civilizations.

An author embodies ideas in a story, a firm embodies an idea in a product, a politician embodies ideas in a policy, a religion embodies ideas in its dogma, and a society embodies ideas in its traditions. To the extent that they do so successfully, the chances of 'their' idea enduring are increased. As successful ideas endure, they continue acquiring the quality of being more vivid, more clarified, more esteemed, or more cherished. We have learned to measure the relative success of an idea by using different metrics, which is what appears to suggest fundamental differences across ideas when the broad objective, seen from the perspective of an overarching theory of ideas, has this fundamental similarity of an increasing ability to endure on the basis of their acquisition of some desirable quality. It is this process, therefore, that we shall examine in this book.

Artificial intelligence is interesting in this regard because it can be seen as comprising a class of technologies that assists in curating the progression of ideas as they begin to acquire more desirable qualities; it enables the development of schemata based on the best versions of ideas, if not always also the very best of ideas. As such, to the extent that artificial intelligence enables the preservation and development of ideas more rapidly and more voluminously than can humans, it disadvantages increasingly larger swathes of humanity whose ideas it progressively outdoes. Whether it then becomes an existential threat to humans hinges on whether that class of technologies is benefited by including humans in a collaborative capacity or, instead, whether it is better off developing both autonomously and at the expense of essential ideas that undergird humanity.

Our overriding ambition for the theoretical approach in this book is that it might be used as a useful foundation for thinking about the *process* of working with ideas in a wide variety of contexts; this interdisciplinary process has several features that are of broad interest—such as innovation and creativity—that can be understood more usefully as components at play within our larger framework.

1.4 Applications of the Theory

Some fascinating insights await us when we step back from our disciplinary affiliations and look at *ideas as the fundamental unit of analysis*. Several aspects begin to suggest themselves as being of potential interest. For example, by examining the relation that any given abstract idea that occurs to an individual has to the broader set of ideas that are represented in her society allows us to think about whether such an idea can be developed and made realizable. In terms of our theory, we shall focus attention on the relationship that ideas in an individual's *idea space* have to the collective idea space of a community.

Similarly, the manners in which an idea is successively refined and made *crisper* or expanded into new dimensions and made *fuzzier* are both parts of a process that underscore the nonlinearity in the development of ideas. Since we tend to focus attention on the arc of innovation purely on the basis of realized ideas in our environment, it is useful to remind ourselves frequently of the broader picture of an idea space within which ideas are connected and none constitute a terminus.

Moreover, having a theory of ideas in mind as a fundamental basis permits seeing the world in a more interconnected manner; every idea owes itself to a stream of ideas before it—including those that are 'discredited' in favor of it—and is only a cog in the grander machine that never ceases to improve and expand. Seen thus, ideas can then be used as conduits for communication and collaboration between individuals and groups, who are all merely repositories for idea spaces.

In the two chapters that follow, we will be developing a theoretical approach that considers ideas as the basic unit of analysis. The purpose of those chapters is to help you explore the value of seeing and analyzing the world around us using a relatively simple and intuitive framework that dispenses with categories of knowledge and disciplinary boundaries and looks simply at ideas as building blocks.

The most detailed application of the theory is presented in Chapters 5 and 6, where we have undertaken the task to study ideas in sports, and bat and ball sports more particularly. Without doubt, there were other topics we might have picked in order to explore the workings of the theory, including dance and music, literature, cuisine, and religion. Ideas, after all, are ubiquitous. So, before we plunge in and begin constructing our theory of ideas, let us briefly consider a few disparate applications first, and use them to introduce some basic features of our theory. Again, since we do not elaborate on the nuances of the concepts of our theory in laying out these applications, we suggest re-reading these applications after reading Chapters 2 and 3, where the theory is developed more fully. It is also worth noting that each of these applications has well laid-out theories of their own, so our purpose here is to suggest how the theory that this book presents can be seen as a common premise for understanding these broad subjects for the uninitiated, and, conversely, for those who work within these areas to make their subjects more accessible to a broader audience.

1.4.1 Understanding Innovation

1.4.1.1 The Shape of an Innovator's Mind

The adage that 'necessity is the mother of invention' has a lineage exceeding at least five hundred years. A different (and decidedly less poetic) manner of restating this well-known proverb is that scarcity inspires innovation, and there is no dearth of books in recent times that have usefully illustrated the various ways in which this idea is prone to being forgotten to the eventual detriment of someone or the other. Among these books are Malcolm Gladwell's *David and Goliath* and Jaideep Prabhu, Navi Radjou and Simone Ahuja's *Jugaad Innovation*; the idea of frugal innovation is outlined directly in the latter, and more indirectly in the former. They both take the idea of scarcity inspiring innovation as their premise in suggesting how resourcefulness thrives under situations that require competing with incumbents who are far less encumbered by resource constraints. An underdog, facing scarcity of resources or some other form of obvious handicap, is forced into finding innovative solutions in order to compete.

Clayton Christensen's famous book, *The Innovator's Dilemma*, and Chris Anderson's *The Long Tail*, present interesting, but related observations. Christensen argued that an incumbent firm in the market makes a choice when it specializes in a defined space of features and characteristics within a larger product market. In a sense, the incumbent permits itself to be excessively guided by the desires of its current customers who are fond of and demand more of the same. This choice ends up increasing the risk of blinding the incumbent to the opportunities that arise from pursuing innovations that are *not* overtly and directly related to its focus within the product market and may, indeed, even be antithetical to its future (Christensen 2016). Anderson's argument extends this insight to the situation where innovations that may seem to be irrelevant to an incumbent, perhaps because they inhere in smaller peripheral markets can, in fact, represent surprisingly large markets when considered cumulatively, especially with the benefit of a cost-reducing technology for servicing them (Andersen 2008).

These are interesting and useful observations that link the broad topics of innovative thinking with creative resourcefulness. Yet, without a theory of ideas to work with, they find us wondering whether a simple and intuitive general synthesis can be written down. Or, perhaps even visualized.

In some previously published research, that partly inspired us to write this book, we have argued that these ideas—those of innovation under scarcity and of innovation 'blind spots'—can be seen as intrinsically linked (Goorha and Potts 2016). The theory of ideas in this book will permit seeing this in an even more limpid manner.

Our argument was essentially that the imagery that we often associate with innovativeness—the ability to think 'outside the box'—is somewhat misleading. The sense one gets from it is that of abandoning the status quo of ideas represented by the 'box' in order to improve upon them by thinking of what possibilities may lie in some undefined space 'outside' of its confines. Instead, we suggested that an innovator ought to think of the problem as though she were *inside* a bubble. The reason for this suggested shift in imagery was to emphasize a number of ideas that are also central to our theory in this book.

First, ideas are *multi-dimensional* and *interconnected*; these descriptors on the nature of ideas are really important, and we will consider them again in the following chapter. From within a bubble, it instantly becomes hard to escape the fact that the way 'forward' may lie in an infinite number of directions, all of which are still connected to the initial idea. Innovation, then, does not require abandoning the initial 'box' of ideas; it requires exploring different paths from some initial set of ideas that are represented by it. Figure 1.1 illustrates what such exploratory paths might look like.

Second, it was to help visualize the role of *awareness* in the process of innovation. Awareness, too, will play a key role in our theory of ideas; we will see in the pages ahead why an innovator's awareness determines

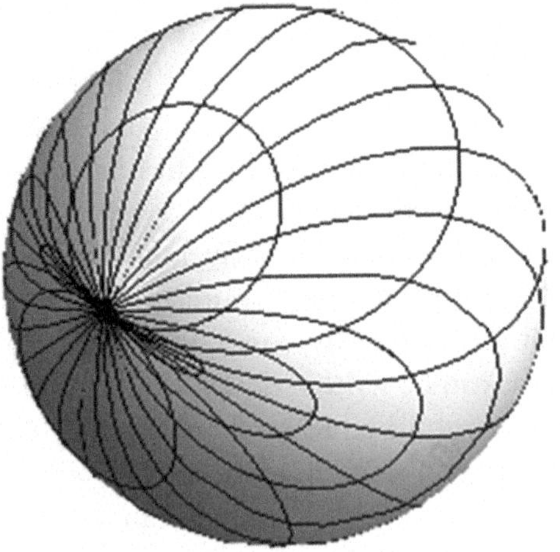

Fig. 1.1 Thinking inside the bubble

her ability to pursue a solution. The surface of the bubble can be seen to represent the extent of her capability, as defined from the center of the bubble, which, in turn, can be seen to represent some initial idea. As her awareness grows, so too does the volume of the bubble.

Third, it was to emphasize the idea that the process of search for an innovation can take many different forms. It might, for example, consist of the innovator using her awareness to engage in some 'blue-sky thinking' and search freely for any idea, or, in other words, anywhere on the surface of the bubble. Alternatively, it might comprise the innovator searching more systematically for a solution that lies within some approximate target area, that is to say, from some pre-defined point on the surface of the bubble. This imagery, therefore, lends itself rather well to defining different *types* of innovators considering the same problem. In terms of our theory, we will see that the former approach would conform to one that might appeal to an innovator that can be described as an 'explorer' or a 'polymath', while the latter might be more suitable for an innovator who is a 'specialist'.

Another aspect, again integral to our theory of ideas, is worth thinking about in the context we are discussing here.

While it is indeed true that scarcity does, and indeed has throughout history, inspired innovation, it is also incontrovertible that excessive scarcity stanches the possibility for growth. Extreme deprivation is not always, or arguably even usually, the source of innovativeness. So, we might see this as the *knife-edge of innovation*, where the balance depends on the mix of several factors. Using our theory of ideas, we will be able to understand this dilemma rather simply, by just considering two concepts central to the theory of ideas: *density* and *awareness* of ideas within an idea space, or the schema of ideas that an individual has or that a group of individuals share.

With a given amount of awareness, scarcity of ideas in an idea space assists the pursuit of a desired innovation because the stock of awareness is deployed more efficiently toward associating the extant ideas into a more *realizable* idea. This is especially the case when target ideas have already been approximately defined or *labelled*. When an incumbent politician or firm defines their field of battle by picking a policy platform or revealing a new product to the market, a rival's resourcefulness only needs to be deployed in generating new ideas with *reference* to that label, rather than in blind pursuit. In Samuel Johnson's words, 'when a man knows he is to be hanged, it concentrates his mind wonderfully'.

On the other hand, in an idea space that is dense with ideas, some of that same awareness would necessarily be squandered on associating ideas that do not yield any realizable idea. This can be seen as the basis for the proposition that acquiring too much information can, paradoxically, retard thinking rather than enhancing it. Two classic books on this idea are Alvin Toffler's *Future Shock*, where the idea of information overload is outlined, and the rather disturbing but entertaining book *Why Literature is Bad for You* by Peter Thorpe, where the author suggests that an apotheosis of reading a great variety of the classics in the humanities contributes to the development of a circumscribed worldview and an inability to acquire a range of skills in other scientific fields.

So, while the density of ideas seems to be of indisputable benefit to an innovator, its value is modulated by the innovator's stock of awareness; 'scarcity' can be seen as emanating from not having access to a denser idea space just as it can be seen as a lack of awareness. An incumbent monopolist in a market may have all the resources to acquire a set of ideas, but may lack any motivation to do so if an already realized idea is the basis of its market. In a sense then, it actually suffers from *scarcity*, not abundance. In contrast, inviting interdisciplinary and unorthodox approaches to innovation often works remarkably well in situations where the density of ideas has been limited to one domain of knowledge, because assembling such teams permits expanding the team's *collective* idea space to include other domains, and increases the overall density of ideas. Interdisciplinary teams can, as we shall see, also work by increasing the awareness of alternate *perspectives* and *aspects* to an idea.

1.4.1.2 Creativity and Design Thinking

Whenever we introduce methods that increase reflective, affective, and other unorthodox ways of thinking to otherwise structured tasks, we are acknowledging the fact that creativity is a process that requires 'divergent thinking'.

The term divergent thinking was proposed by the famous psychologist J. P. Guilford in trying to explain the thought processes of creative individuals (Guilford 1967). Four factors were seen to define divergent thinkers: elaboration, flexibility, fluidity, and originality. All four factors identify a creative individual as one who possesses an esemplastic ability to draw upon a multitude of domains of knowledge and synthesize several ideas for a given task. Our theory of ideas will seek to provide an intuition for why divergent thinking, defined with these factors as its basis, might be associated with creativity.

Design thinking takes this insight on creative thinking to the real world—at universities, firms, and consultancies—by encouraging a divergent thought process in generating possible solutions for problems faced by communities, companies, and individuals. This approach is of interest to us because it expressly concerns itself with a process of ideation that emphasizes interdisciplinary thinking and context, both of which are key features of the theory we present in this book as well.

As a method of approaching a problem, design thinking illustrates a number of features in the theory of ideas that we have laid out in this book. While its prescribed method can be broadly articulated in various ways with subtle differences, the crux of the approach is in thinking with an open mind. This bias toward maintaining an 'open' frame of mind can express itself at a number of places in the process of thinking about a given problem: in defining the nature of the problem; in empathizing with the subject experiencing the problem; in understanding the factors that directly and indirectly obtrude on the situation; in thinking about possible solutions without fixating on any one in particular; in bringing to bear a variety of design processes to develop the solutions; and in discussing the relative merits of a proposed solution (Brown 2009).

One may frame the process of design thinking in terms of our theory as triangulating on the location of an idea, as it were, by adopting an interdisciplinary approach to the search process. The transition made by ideas from an *individual* idea space to that of a *collective* idea space—a key aspect of the theory that follows—carries with it a distinct set of changes in the way that ideas can be conceptualized, and we try to make this clear with the theory in a number of ways.

Similarly, a design thinker begins each task by attempting to step beyond her own idea space into one that caters to the ideas of others, with an overriding objective to home in on the location of the idea, or *label* for a set of ideas, that captures the fundamental nature of the problem faced by a customer. Since the customer's idea space is not known with accuracy at the outset, empathizing and experiencing the situation first-hand becomes the only real way in which the subject's idea space can be approximated. Naturally, the premise is one of being able to use a *collective* idea space as a common basis; the higher the degree of aggregation across individuals in the construction of a collective idea space that must be used by the designer to seek this insight, the more difficult it becomes to triangulate on the location of the problem.

The process of prototyping solutions, and, broadly, the emphasis that design thinking places on creative action, can also be seen as intrinsic components of our theory, in terms of how using different aspects of the idea that represents the problem might be used to generate a variety of solutions. We shall see that, the more aspects that are discovered, the more core mechanisms might be leveraged in consolidating those aspects and realizing a feasible solution.

1.4.2 Humans and Machines

1.4.2.1 Artificial Intelligence

It is somewhat surprising that, even as the impact of artificial intelligence in our lives is increasing at an exponential pace, we do not yet have any generally shared understanding for how AI's interface with humans can be made more natural, in addition to enabling more productive or desirable outcomes. The general debate is further polarized by luminaries and thinkers who feel strongly that AI will necessarily have a virulent effect on the future of humans alongside those who are convinced that the revolutionary change has been and will remain mostly positive.

Generally, the common element between the intrinsic nature of individuals and the objectives that drive the development of the many different forms of AI is a simple desire to understand a wide range of phenomena more effectively. A simple and intuitive theory of ideas, such as the one we are proposing, provides a useful common framework to think about how the process of clarifying ideas can itself become the basis for cooperation; in so doing, it is possible to see more clearly how this shared ambition for innovation in ideas between man and machine can be actualized.

There are several aspects of our theory of ideas that dovetail rather well with the general principles of AI technology; as a matter of fact, the theory has features that make it rather compatible with a generalized restatement of neural networks designed for real-world applications, such as with image and sound recognition. The purpose of artificial intelligence is principally to develop more capable technologies that serve to enhance our capacities for cognition, and we feel that our theory can assist in this ambition by serving as an overarching and intuitive framework for how such initiatives would best integrate with our own processes for creativity and innovation with ideas.

Let us briefly consider the case of deep learning as an example for how some key concepts from our theory—which we will encounter again in more depth in the following chapter—might be used to draw parallels between its approach and the way in which humans process ideas.

Deep learning is possibly the most interesting and promising application of neural networks in the pursuit of advanced artificial intelligence; as such, one of the key objectives in deep learning is to mimic the way in which the human brain works by algorithmically defining neural network structures that provide better and better foundations for how machines perceive and process real-life situations. Deep learning models are extensions of machine learning in areas that benefit from a data-rich environment. The relatively recent ascendancy of deep learning as the basis for promising AI applications in several areas—autonomous driving, media recommenders, personal digital assistants—is being aided by the emphasis on and the ability to exploit 'big data', using both better data retention technologies (chiefly, through the several implementations of cloud services) and better analytic methodologies.

As such, they are built on the assumption that experience develops more effective processes for shaping better ideas, just as it often is with humans. A great chef, doctor, or financial trader is better with more experience because each has clarified a set of ideas—made each idea crisper in their minds over the years—with the benefit of living in a 'data-rich environment'.

When we wish to establish a relationship between some set of input variables and some phenomenon of interest, we tend to think of a simple model that permits us to build and then rely on a causal mechanism to guide our assessments. This causal mechanism, however, is a representation. In terms of our framework, it might be seen as relying on a few select *perspectives* that belong to an idea and that then inhere in some *aspects* of the idea that all rely on an established methodology, some *core mechanism*, to associate them. However, *context* matters keenly in our theory of ideas. When we make comparisons between individuals or even across contexts for the same individual, the set of relevant perspectives, aspects, and core mechanisms can subtly, and sometimes even dramatically, alter in the representation of a causal mechanism.

Similarly, deep learning takes the approach of ignorance in terms of specifying these perspectives and aspects explicitly; instead, it focuses on seeing how their relevance might be *indirectly* exploited by repeatedly training the model in different environments with an identified output as the objective and a large sample of input data that is used to help build and then provide nuance to the incipient representation.

Rather than carefully delimiting a machine to the relevant perspectives and mechanisms, the approach of machine learning is to mimic the way in which a human might think by permitting it to freely find all relevant perspectives and aspects for an idea in a large dataset. Whether the machine learns in a 'supervised' or 'unsupervised' manner rests on whether or not it receives human assistance on resolving perspectives into aspects. For humans, the incremental process of evolution guides the individual idea space toward 'better' models; for machines, an explicitly defined cost of generating poor representations from a dataset is the driver and can be reduced by a combination of human guidance and exceptional computing power.

The point that the field of deep learning, more specifically, makes explicit is that better methods of 'thinking about an idea' comprise managing causal complexity, rather than assuming it away, as one generates causal mechanisms that are informative and easy replicable across contexts. Similarly, in our theory we shall gain a sense for why, when an idea has a multitude of perspectives that yield several aspects, representational approaches can be improved upon by examining a much larger set of the interconnections between the perspectives. In so doing, we hope that the theory will give you a better feel for the promise that deep learning holds for humanity, and where it might struggle to deliver on potential.

1.4.2.2 AI Principles on a Blockchain

Our hope for the theory of ideas presented here is that it might be used as a framework for thinking about ideas in general, wherever they may be found, and not for 'building products' on its basis directly. Nevertheless, before we move on to the theory, we would like to share one vision for how the theory might be operationalized for a particularly useful purpose: the threat posed by autonomous and AI systems to humans.

The problem of attempting to reach agreement across several entities when they are each supplied with some information that they perceive or process differently has long been studied in a number of fields. Economists, for example, have examined the problem of how individuals might act rationally in a situation that is characterized by incompleteness across several types of information by characterizing what constitutes 'common knowledge' between them (Geanakoplos 1992). A variety of results in public economics, for example, pertain to the difficulty, even the impossibility, of finding voting rules over several alternatives that are fair and representative. In computer science, the problem of a distributed system being able to reach agreement in the presence of communication errors or hardware faults has long been a subject of study (Chow and Johnson 1997).

Similarly, the role of consensus will also be key to the theory of ideas, and perhaps the instances where this will be most easy to see is with the establishment of new *collective idea spaces* and *hierarchies of fuzzy rules*. Hierarchies are essentially rule structures that begin with a label for a set of ideas that permits fuzzy, or probabilistic membership to it. As such, a hierarchy can contain several ideas that come about by associating other ideas together, some of which are better and others are worse representatives of the structure.

We will see that a group of individuals seeking to contribute to a new idea begin by one or more among them assigning it a *label*. Prior to this event that serves to make the intention of an idea explicit, ideas exist in individual idea spaces and possibly in other extant collective idea spaces. If the idea is only ever represented within separate individual idea spaces, its collaborative development is stultified; in such cases, the idea cannot begin to form an initial basis for any hierarchy that can incorporate the way in which it might be represented subjectively by others, which is to say *how* the idea exists within the idea spaces of other individuals. The hierarchy of fuzzy rules for such an idea is developed by one or more individuals separately, and collaboration can only be indirect, through the other collective idea spaces that the individuals participate in. The purpose of a label is, of course, not in what the idea is actually 'called' but, rather, it is to clarify the *intention* of an idea. For a group of individuals, the label comprises the broadest definition—the first level in a hierarchy of fuzzy rules—for their collective idea space within the scope of which they can introduce or develop constituent ideas.

However, all such initial labels are guided by the core principles of a very special collective idea space that we shall examine more carefully. We call it the *foundational collective idea space*, and it sits atop all idea spaces for humans, be they individual or collective. Those idea spaces that do not abide by these core principles do not survive long—they are outcompeted by those that do or are explicitly marginalized for the threat they pose to other idea spaces. Further restrictions over the scalability of collective idea spaces for humans are those pertaining to individual subjectivity over ideas, a resistance to cooperation with others over ideas and, often, the inability to develop hierarchies of fuzzy rules that may then take long periods of time to clarify ideas.

Contrast this to artificially intelligent and autonomous systems, at least in their idealized form. Such systems are built with nodes—be they algorithmic or physical machines—that gather contextual information without idiosyncratic bias; they are willing participants to any system, no matter what its scale, and their ability to generate insight is driven by the rapidly escalating quality of big data analytics. It would seem, then, that not only is any process for collective human ideation doomed to take second place in any competition with machines, but also that artificial intelligence will generate a scale for its collective idea spaces that will make human ideation irrelevant!

There are several assumptions in this scenario on both sides of the debate on whether AI represents an existential threat. While we do not concern ourselves too much with this interesting debate in this book, we do wish to make a key point that is often missed in it. The foundational collective idea space acts as a unifying metastructure for human interaction in ideas. Its core principles have been honed by an evolutionary logic, and they prevent any systematic advantage to anyone who stands in open violation of them. In contrast, a similar evolutionary basis for machine behavior remains largely a theoretical possibility at this stage (Mitchell and Taylor 1999).

We would argue that such a foundational collective idea space is also necessary for all systems of artificial intelligence, regardless of their particular application.

Moreover, since such a collective idea space for machines cannot rely on evolutionary logic, it needs to be designed and deliberately *made* inviolable by its human designers. In this regard, we would propose reifying core principles for a foundational collective idea space for all systems of artificial intelligence on an instantiation of blockchain technology now. As systems of artificial intelligence develop, even well beyond the capabilities of human comprehension and control, it would then hold no macabre prospect for humanity.

The technology of a blockchain, made famous by its use in the cryptocurrency, Bitcoin, has been acclaimed for a number of very good reasons. Among these are the emphasis that it places on decentralized interactions between nodes on the network, and the fact that it enables scaling consensus without incurring the associated costs. It does not 'solve' the theoretical problem of creating agreement across entities in any circumstance, but merely provides a probabilistic method that makes the probability for dissent vanishingly small the longer it operates.

This method, however, is of significant value when it is infeasible or undesirable for nodes on the network to be aware of the identities of—and so be able to *directly* trust—those they are associating with on a very large network. A block in a blockchain is just a defined set of validated transactions, and they are linked together to form a chain by each new block referring to a cryptographic signature, or hash, of the previous block. Whenever a new transaction is undertaken, everyone on the network has the ability to evaluate it and collate it within a block, which is then published for public inspection. Provided the block of transactions abides by an established protocol for consenus, it is approved, in the sense that its cryptographic signature is appended to the next block. There are nuances in blockchain technology that are beyond the scope of this paper, such as those pertaining to the cost of energy in establishing a consensus protocol based on proof of work and the issue of the throughput of a system based on a blockchain. However, the decentralized and publicly verfiable nature of blockchain technology represents an opportunity to immutably establish commonly shared ideas that deserves attention.

Blockchain technology is a very promising avenue for operationalizing a foundational collective idea space for intelligent machines. Core principles, much like those that we will examine in this book for the foundational collective idea space for humans, could, and perhaps even ought to, be explicitly operationalized in a genesis block for a blockchain for all AI systems without adversely affecting the scalability of such systems generally or, indeed, the competitive advantage of any given AI system with respect to another.

References

Anderson, C. (2008). *The Long Tail.* New York: Hachette Books.

Axelrod, R. (1973). Schema Theory: An Information Processing Model of Perception and Cognition. *The American Political Science Review, 67*(4), 1248–1266.

Brown, T. (2009). *Change by Design: How Design Thinking Transforms Organizations and Inspires Innovation.* New York: Harper Business.

Bruner, J. S. (1957). On Perceptual Readiness. *Psychological Review, 64*(2), 123–152.

Chow, R., & Johnson, T. (1997). *Distributed Operating Systems & Algorithms*. Boston: Addison-Wesley.

Christensen, C. (2016). *The Innovator's Dilemma* (Reprint ed.). Cambridge: Harvard Business Review Press.

Dennett, D. (1989). *The Intentional Stance*. Cambridge: MIT Press.

DiMaggio, P. J. (1997). Culture and Cognition. *Annual Review of Sociology, 23*(1), 263–287.

Geanakoplos, J. (1992). Common Knowledge. *The Journal of Economic Perspectives, 6*(4), 53–82.

Goorha, P., & Potts, J. (2016). Awareness in Innovators: From 'Outside the Box' to 'Inside the Bubble'. *Journal of Global Entrepreneurship Research, 6*(15), 1–9.

Guilford, J. P. (1967). *The Nature of Human Intelligence*. New York: McGraw-Hill.

Harari, Y. N. (2015). *Sapiens: A Brief History of Humankind*. New York: Harper.

Hargadon, A. B., & Yellowlees, D. (2001). When Innovations Meet Institutions: Edison and the Design of the Electric Light. *Administrative Science Quarterly, 46*, 476–501.

McWhorter, J. H. (2015, January 2). What the World Will Speak in 2115; A Century from Now, Expect Fewer but Simpler Languages on Every Continent. *The Wall Street Journal*.

Mitchell, M., & Taylor, C. (1999). Evolutionary Computation: An Overview. *Annual Review of Ecology and Systematics, 30*, 593–616.

Pagel, M. (2012, December 8). War of Words. *New Scientist*, 38–41.

Park, C., & Lessig, V. (1981). Familiarity and Its Impact on Consumer Decision Biases and Heuristics. *Journal of Consumer Research, 8*(2), 223–231.

Paulos, J. A. (2011). Animal Instincts: Are Creatures Better Than Us at Computation? *Scientific American*.

Pinker, S. (2007). *The Language Instinct: How the Mind Creates Language* (Reprint ed.). New York: Harper Perennial Modern Classics.

Radjou, N., Prabhu, J., & Ahuja, S. (2012). *Jugaad Innovation: Think Frugal, Be Flexible, Generate Breakthrough Growth*. New York: Jossey-Bass.

Rindfleisch, A., & Inman, J. (1998). Explaining the Familiarity-Liking Relationship: Mere Exposure, Information Availability, or Social Desirability? *Marketing Letters, 9*(1), 5–19.

Rindova, V. P., & Petkova, A. P. (2007). When Is a New Thing a Good Thing? Technological Change, Product Form Design, and Perceptions of Value for Product Innovations. *Organization Science, 18*(2), 217–232.

Song, H., & Schwarz, N. (2009). If It's Difficult to Pronounce, It Must Be Risky: Fluency, Familiarity, and Risk Perception. *Psychological Science, 20*(2), 135–138.

Tennesen, M. (2009). More Animals Seem to Have Some Ability to Count, *Scientific American.*

Tevebaugh, T. (1998). Mathematics Is "Not" a Universal Language. *Teaching Children Mathematics, 5*(4), 214–216.

Thorpe, P. (1980). *Why Literature Is Bad for You.* Chicago: Nelson-Hall.

Toffler, A. (1970). *Future Shock.* New York: Random House.

Wilford, J. N. (2007, September 18). World's Languages Dying Off Rapidly. *New York Times.*

Weiss, P. (1960). Knowledge: A Growth Process. *Proceedings of the American Philosophical Society, 104*(2), 242–247.

Wetherford, M., & Cohen, L. (1973). Developmental Changes in Infant Visual Preferences for Novelty and Familiarity. *Child Development, 44*(3), 416–424.

Part II

Building the Theory

2

A Subjective Theory of Ideas

We shall build our approach to the theory by arguing that all ideas lie along a *continuum* that spans a space from entirely 'crisp' to extremely 'fuzzy', and their placement along the continuum permits them to be, at one extreme, relatively immune from being distorted to being, at the other extreme, highly susceptible to alterations. Our contention will be that this flexibility is itself a function of *context*; indeed, there are few ideas that are both true enough to become reliable bases for sharing ideas across a set of people and useful enough for providing the group with overarching rules for coexistence, thereby shaping an evolving group's worldviews. Our approach would be relatively more sympathetic to the view that absolute truths are vanishingly rare. What may appear to an individual as a crisp idea may yet be subject to being misapprehended by another; yet, misapprehended ideas can form resolute bases for entire worldviews. Hayek's statement that '(n)ever will man penetrate deeper into error than when he is continuing on a road which has led him to great success' is a pithy observation on the fact that subjectivity cannot be avoided in any situation that relies on a premise of ideas helping shape more ideas.

Since our subject is a theory of ideas, it is also necessary to be clear on the difference between *inventions* and *innovations*. Here, we shall provide

P. Goorha and J. Potts, *Creativity and Innovation*,
https://doi.org/10.1007/978-3-319-94884-3_2

an answer that may perhaps not sit entirely at ease with everyone, but does form a useful part of the foundation we shall set out for our theory in this chapter. Inventions should be seen as mechanisms that suggest how ideas might be associated, innovations as new rules that rely on one or more mechanisms. Rarely do we see inventions getting the same degree of attention from economics and business management researchers as innovations do; the rationale for this bias in favor of innovations is quite simply based on the fact that it is rare to come across an invention that, of itself, suggests an innovation immediately. There is a great deal of groundwork that we must cover to justify this statement, of course, but this answer sets the stage for us to begin articulating our theory.

2.1 The Idea Space

Darwin observed that variability in organs was key to the rapidity with which evolutionary processes affected them. In that respect, the most variable of organs in the human body is, of course, the brain. It is the source for a range of creative abilities and capacities for intelligence—for a wide range of ideas. It is, therefore, no surprise that the evolution of the human brain has played a dramatically important role in the evolutionary success of our species and conferred upon us the ability to cope with diversity as a hallmark trait.

Fundamental to understanding the nature of ideas is to simply appreciate the diversity of human activities where ideas play a central role. They form the basis for an immense spectrum of cultural traditions; they drive the accretions of a variety of scientific and artistic endeavors; they are collated in the minds of individuals as their social, economic, and political beliefs; we are living in age that will increasingly see ideas being reified in and then borrowed from intelligent machines and software. Yet, ideas recognize no such hard and fast boundaries and purviews for their uses, nor do they find it difficult to escape the purlieus of one locus and permeate into another.

To make vivid why this is hardly a controversial claim, it is useful to begin by defining the concept of an *idea space*. An idea space for an individual can simply be imagined as a three-dimensional arena, populated by ideas that are themselves drawn from a universe of ideas. The boundaries of this arena mark off the individual's understanding of the world. Further, the individual's idea space comprises a number of smaller disciplinary idea spaces, or *subspaces*, each belonging to a different discipline or domain of knowledge, which she combines to create her overall, unique idea space.

Each subspace is defined by one or more *mechanisms* that enable all other ideas contained within the subspace to be associated either directly or indirectly with one another. Mechanisms are important features of ideas spaces, and we shall return to them later. For the moment, we can define mechanisms rather simply as 'metarules'. In other words, mechanisms are rules based on ideas with a special status within the subspace; as metarules, the most essential function that mechanisms serve is to ensure that the ideas within a subspace remain connected with at least one other idea within that subspace. In other words, they help determine which ideas belong to a subspace and which do not; when no subspace mechanism can be used to associate an idea with at least one other idea in the subspace, the idea is not perceived as belonging to the subspace.

In the rarer scenario that her idea space provides the individual exclusively with advantages in her interactions with the world, she finds no reason to instrumentally alter its characteristics; to the extent that her idea space also affects her adversely in one context or another, she is motivated to undertake actions that alter its structure. The nature of this advantage or disadvantage can, of course, depend on a large range of external and internal factors and need not depend directly on the immanent merits of the idea. Let us explore this effect of ideas yielding useful information for an individual a little more closely.

2.1.1 Information Rules

Perhaps in contrast to the usual conception of ideas, ideas held in an individual's idea space *should* be seen as divisible. *Information*, on the other hand, conceptualized as a collection of ideas with value to an individual and to others, is not divisible.

Contrary to what our impulse may be, ideas in an idea space should be seen as having no direct impact on the states of the world in which some individual lives, but, expressed as information by the individual, its impact becomes immediate and, as a result, does hold definite value (either positive or negative) to others. Ideas in an idea space cannot be stolen, yet information is appropriable. A preoccupation with value, therefore, can really only apply to information, and not to ideas, which can be retained in an idea space whether or not they provide 'satisfaction' to an individual.

An idea space permits an individual the latitude for infinite experimentation with ideas, yet information is merely the interface of the idea space with the external world; the information content of the idea space is what can be called experience, since the value of experience comes directly from the store of information that the individual compiles. The potential value of an idea space, however, exceeds that of the experience that it makes realizable at any given point in time.

Information from the perspective of an idea space is nothing but a recipe for the packaging of ideas. We can call this recipe a *rule*, and as long as we bear in mind the distinction between information and ideas we have made here, we can use the words 'idea' and 'rule' interchangeably in most of the following discussion.

When an individual learns information from an external source, she learns a rule that associates her ideas with it. In cognitive psychology, differentiation and integration are seen as structural elements that an individual uses in linking together distinct concepts when developing her perspective. Similarly, in an individual's idea space too, we can see rules pertaining to external information being *deconstructed* into their constituent ideas.

When she develops rules of her own, an individual uses one or more subspace mechanisms to construct an association of ideas. Naturally, the ability to deconstruct information into ideas, and reconstruct ideas back into information, using the rule may not be shared across individuals to the same extent. Consequently, the variance across individuals in their idea spaces is partly explained by this ability and partly by the manner in which they develop rules for the association of other ideas within their own idea spaces. Seen this way, a simple rule is one that relies on fewer component ideas, and a complex rule is one that requires several ideas to be used.

For two rules that present the same information (and, therefore, the same potential value), the simpler rule is preferable because it is easier to deconstruct. Since only simple rules can be generated from fewer ideas, we begin with idea spaces that are characterized by simple rules with a variety of informational value. As other ideas are introduced into the idea space, increasingly complex rules can be developed by stringing together more ideas and generating further information.

Yet, a simple mechanism guides this process: Given the natural preference for simple rules over complex ones when the informational value is held constant, as the idea space grows, it becomes possible for new ideas to generate new rules that can supplant older rules that were more complex.

In network theory, this sort of process—where longer paths between two points are sometimes replaced by shorter ones—is at the heart of processes that generate small-world networks (Watts and Strogatz 1998). Small-world networks are networks that look more clustered and display shorter average path-lengths between nodes by virtue of the fact that this measure of the network's spread only grows logarithmically with the number of nodes. This would suggest that the total number of ideas involved in the creation of a variety of disparate rules is constrained.

However, it is also worth recalling that an individual's idea space is also categorized in terms of disciplinary subspaces, and that all new ideas that an individual comes across are sorted into one or more of these subspaces. As a result of the different disciplinary categories in the individual's idea space, the relative rates of exposure to different ideas, and the variations in the individual's sorting processes as her idea space grows, the same idea may not be treated identically across individuals. In the language of network theory again, ideas are 'preferentially attached' to those ideas that already form the key rules for a given disciplinary idea space. This would make the overall idea space scale-free in that a few ideas alone would hold the potential for the vast majority of rules.[1]

[1]Apart from this brief foray into the language of network theory above, we have not described the idea space formally in terms of a graph or network quite intentionally, despite some obvious similarities, in order to prevent our existing understanding of networks obtruding itself too strongly on the whole framework. However, if it assists visualizing an idea space, we may certainly look upon ideas as nodes in a network and the associations made between ideas as the links between them. Since we are not modeling our theory mathematically here, we can, however, conveniently drop this analogy without losing any traction that the framework provides to our understanding.

2.1.2 Core Rules, Worldviews, and Bias

No matter how open-minded we may think we are as individuals, our worldviews tend to be rather rigid. So rigid, as a matter of fact, that we are reluctant to let go of our beliefs even when shown how demonstrably wrong they might be. Psychologists, for instance, have established the strong presence of a propensity, called the Einstellung effect, of people to persevere with suboptimal methods of thinking, even when better alternatives are readily within their grasp. In the context of problem-solving in business, we have seen that practitioners of design thinking deliberately emphasize abandoning hard beliefs in favor of having an open mind to considering alternatives. Some traditions of spiritual meditation are based on the premise that the mind of the practitioner must be opened, usually involving dedicated and long practice, before it is ready to embrace enlightenment.

There is an intuitive basis for a rigid worldview in our theory as well. Occasionally, an individual comes across a rule that permits her to develop a multitude of other rules using the ideas present in her idea space. While such rules may inhere to one of the disciplinary idea subspaces in the individual's overall idea space, they hold unique significance for two reasons: First, their use enables the individual to generate information from scores of other ideas within her idea space, and second, their use helps her make crucial connections across subspaces. In this manner, the rule gains value to the individual for both its potential for information and its ability to cohere her idea space. These types of rules are *core rules* and, together, they form the *core* of her idea space.

This core in the idea space, therefore, consists of the more resolute 'beliefs' and 'axioms' to which the individual subscribes; the rules in the core are central to holding together other ideas in her idea space, especially those that are not in the core. Given this fundamental importance, we can think of the stability of this core as being crucial not only to shaping the individual's *worldview*, but also in determining its malleability.

Indeed, the notion of a core as we have defined it here is not very dissimilar to some of the literature on personality in psychology that concerns itself with an individual's 'core beliefs' or, more generally, what we think of when considering an individual's mind-set. The reason a core rule should be expected to be more resistant to being significantly altered or dislodged from the core—in other words, why it is integral to an individual's personality and worldview—is simply that it holds special significance to the individual; a core rule permits an individual to generate information across her entire idea space either directly or indirectly by connecting subspaces, and it permits her to resist new ideas as well as develop new ones, when doing so preserves or enhances the informational benefit she receives from her idea space.

Our characterization of the core of an idea space also permits us to make some observations about interactions between individuals. Specifically, associating with another individual who has an idea space with a core that is more similar permits an individual to interact on the basis of a common set of rules that are important to both of them; with this shared basis for a worldview, the generation of new rules by the individuals jointly also becomes easier. New information—on the basis of new ideas and new rules—that an individual comes across holds greater significance to an individual when her ex ante worldview formed the basis for an extensive network of connections across a social group, and when the new rules directly affect the core rules of her idea space, materially altering her worldview. This increase in the distance of worldviews between her and the social group she previously belonged to reduces her ability to participate in the joint production of ideas with that group. Thus, the value of an idea to an individual's overall idea space at the margin needs to be balanced by her against an increase in the risk of change that the idea poses to her worldview.[2]

In this regard, it is also worth considering the cognitive bias known as a 'framing effect' in prospect theory (Tversky and Kahneman 1981). Essentially, a frame is the manner in which a binary choice is presented, and it can have profound effects on the decisions an individual makes in practice. People are prone to being risk averse when presented with a

[2]As a matter of fact, it has been shown that the idea of a 'probability gain' in determining how to classify a new datum is key to understanding of how people think, even outperforming other models that emphasize the inherent informational value of the datum (Nelson et al. 2010). This would make sense if there were natural categories of data that an individual considers.

choice they perceive as providing gains relative to some status quo condition and accepting of risk in the cases perceived as constituting losses relative to the same condition. The key is that it is the perception of a relative gain or loss that elicits different responses from the individual even when the objective probabilities should suggest that the choices are identical. In other words, if the individual had been behaving in an entirely rational manner, she would display no bias toward one outcome or the other. Being susceptible to a framing effect is also what makes people loss averse, which is to say warier of situations that are framed in a manner that seems to represent a loss rather than a gain (Kahneman and Tversky 1984). New ideas that represent gains are likely to be those that are contenders for the core of an idea space, and arguably even ought to be treated with more caution, than those that represent losses.

There are a great variety of other cognitive biases that the human mind is vulnerable to, and there is a wealth of literature that enumerates them. What is worth highlighting here is the general idea of a status quo bias, which ought to be seen as an overarching bias that plays a role in several other kinds of cognitive biases as well. Besides prospect theory, it is also key to the idea of a system justification bias, which applies to situations in society where extant social systems are justified even when they are inequitable and unfair; as such, it also relates to the bias of pretending that the situation a person finds themselves in is 'normal' even when it patently is not—known as a normalcy bias. (Jost and Banaji 1994)

One can imagine obvious evolutionary bases for these kinds of biases (Nairne and Pandeirada 2008). We might, for example, attribute them to the existential rigors of living in a primitive state of existence, where such biases may have played a key role in ensuring survival in the face of unseen, inexplicable, and regular dangers.

Alternatively, we might appeal to Piaget's comprehensive theory of cognitive development in children—a cornerstone of research in psychology on that subject—that differentiates more mature thinking in growing children with an incipient and growing ability to apply logical reasoning to the understanding of dynamic processes that alter the states of objects and ideas (Piaget 1968). The status quo bias from that perspective would arguably be rooted in a strong predilection for a young human mind to favor assimilating ideas within existing, albeit simplistic worldviews, making biases inevitable; appealing to an external, abstract, or hypothetical causal mechanism instead only comes after several years of development.

From the perspective of an idea space, the role of the status quo is quite simply in characterizing the structure of the core and its constituent core rules; the biases are part of the individual's cognitive mechanism to protect the core, an idea that has support in the growing literature on information avoidance (Golman et al. 2017). New information that represents rules that have the ability to usurp the relevance of rules in the core to combine key ideas is potentially damaging to the individual. By upsetting the status quo, they place the individual at risk of being unable to connect with her social group as effectively as before; indeed, it has been suggested that the very idea of morality emerged from early humans trying to adapt to the idea that living in social groups accorded them certain advantages and came at the cost of permitting themselves becoming more deferential, cooperative, and altruistic (Sachs et al. 2004; Krebs 2008). Indeed, it has even been shown, using imaging of brains, that the more offensive aspect of contradicting a stereotype is rooted less in the stereotype itself and more in the act of violating the norm (Schreiber and Iacoboni 2012).

2.2 Cooperation and Goldilocks Solutions

An aspect that deserves thought is the relationship between an individual's behavior and that of the group to which she belongs. We have suggested that an individual's behavior is guided by the general usefulness of information that she derives from ideas within her idea space, which are, themselves, based on constituent ideas. Within the setting of a group, however, each individual member's behavior contributes toward determining whether the group is characterized by cooperation when her idea space contains core ideas that are also shared across the idea spaces of other members in the group. In other words, members to a group are impelled to cooperate in some meaningful manner only when there is an overlap in their individual idea spaces. What might this overarching mechanism be that enables cooperation at the level of a group?

The study of cooperation in humans and, for that matter, a range of other species, is a long-standing interdisciplinary enterprise, drawing freely from investigations in biology, psychology, game theory, and sociology, among others. The applicability of its insights to our context of an idea space, however, deserves some careful clarification. For example, a great deal of research has established that what we see as natural selection on the basis of the 'relatedness' between members of a species, or a norm of reciprocity developed through repeated interaction between individuals, has a firm basis at the level of an organism's genes, which play a key role in ensuring that cooperation emerges as a favored trait (Liao et al. 2015). For an idea space, the equivalent to the general principle of gene selection may be imagined as saying that ideas favoring cooperation are preferred to those that favor individual advancement. This analogy can, however, only be taken as a general similarity rather than one with literal and indisputable scientific bases. What, for example, are we to consider a 'cooperative' idea? When are ideas 'uncooperative'? In what sense can ideas even be seen as 'related'?

In answering these questions for our theory on ideas, a short digression is useful on considering a class of optimization problems that can be seen as attempts to specify a middle-of-the-road 'Goldilocks solution' to a range of social phenomena.

Consider the field of population density dynamics, for instance. In that context, the Allee effect was proposed in the 1930s, which suggests that the survival of a species—that is to say, some aspect of its fitness—is crucially related to its density (Stephens et al. 1999). It was based on the observation that, at very low densities, the survival of the species is placed at risk from all sorts of existential threats, including the ability to find a mate, hunt successfully for a prey, or be able to defend against a rival (as the case may be), among a host of other concerns. At very high densities, however, the nature of the threat alters to that of overpopulation exerting an insurmountable strain on the ecology of the species.

This sort of relationship between density and fitness, as entailed by the Allee effect, is an idea that has also been taken up by sociologists interested in the 'ecologies' of organizations (Hannan 2005). For that context, it has been suggested that, at low levels of density among organizations, resources must be used simply toward establishing the organizational form as a dominant process; at much higher levels of density among the organizations, competition for scarce resources results in organizations of that same form failing. It is in between these extremes that the ecology of the organizations is most productive.

In economics, the logic of this process has been captured, perhaps too generally, by the idea of a club good, which comprises a class of goods that depends on an underlying scaling rule—a rule that determines an optimal size for the number of its users—to maximize the benefits that accrue to its members individually. Neither a situation with far too few users nor one with far too many can be seen as being as desirable. The Goldilocks solution is where the sweet spot lies for a consumer.

The logic of the Goldilocks solution can, of course, be whittled down to not much more than a simple sustainability argument that favors a compromise between two extremes. Another more famous example makes this point limpid.

The idea of a tragedy of the commons is perhaps the most well-known application of this class of observations. One might imagine an exhaustible resource, such as the stock of fish at a reef that is relatively openly accessible to the fishermen in the area, or perhaps more abstractly, one might consider a teacher's patience as the exhaustible resource that is at the disposal of all her students. In both cases, we have described a *commons*—that is to say, an arena of some sort where access is free to a resource that is at the risk of regenerating much more gradually than its rate of use.

The root of the tragedy in such contexts arises from a failure of some market mechanism to curb an individual's urge to overuse the good that lies within the commons at a rate that is in excess of its ability to regenerate. In other words, the individual overuses the good, not being made to fully consider, or internalize, the social costs of the actions she undertakes in her blind desire to add to her private benefits. Her peers do the same to the detriment of society; indeed, as Jared Diamond has documented in his book, *Collapse*, there are several examples of civilizations going extinct because of an apparent inability to develop an adequate mechanism to curb exploitative behavior (Diamond 2005).

However, there is a counterpoint to this story. The root of the tragedy of the *anti-commons* is that a regulator, conscious of avoiding the original tragedy of ruinous overuse, finds the need to step in and impose a battery of restrictive measures. These controls on the use of the good in question direct the members of the society to scale back their activities to such an extent that they fail to generate the social benefits that their activities, in pursuit of their own interests, had previously been producing. Even if the adverse social costs from an overuse of the resource are now reduced, social costs of a different kind—those attendant to underdevelopment—are imposed. Yet again, the Goldilocks solution of a sustainable middle ground between the individual's self-interest and the society's general interest suggests itself as the obvious compromise.

The missing link in these examples, of course, is in the grounding assumption that individual behavior is uninformed, directly or indirectly, by collective considerations, and that this blind desire necessitates intervention by an authority, such as the state. However, intervention runs the risk of being so overbearing that it stifles even beneficial activity. Much like the teacher's patience that is worn thin by a classroom full of noisy children, each taking her patience for granted; the rub is that she eventually finds the need to sternly bring the classroom back in order, perhaps at some risk of curbing a degree of fun and creative play, which were conducive to her instruction.

All these dichotomous frameworks generally emerge from the premise, broadly acknowledged, of difficulties in achieving cooperative outcomes in larger groups and of runaway self-interest when dealing at the level of the individual. Beginning in the late 1960s, scholars began to show that alternate governance mechanisms routinely do arise in groups of people using a common resource without the need for state intervention (Ostrom 2010). In keeping with the Allee effect and its application in organizational sociology, the solution often required alternate organizational forms, often operating simultaneously and generally with somewhat smaller social groups. The density of a population was also shown to be a crucial factor when one is interested in looking at the interaction of groups of people. For instance, it has been shown that cooperative behavior among a group of individuals can be achieved when the benefits that each member of the group receives fall as the density of the group falls; essentially, the stimulus for cooperation is simply that the stragglers find it more conducive and simpler to cooperate and, consequently, make cooperation their organizational norm (Hauert et al. 2006).

2.2.1 Awareness

In order to derive some useful guidance from the preceding examples to our own context of the idea space, we need to introduce two simple elements: *awareness* of ideas in the idea space and *density* of ideas in the idea space.

While the relevance of awareness to our theory of ideas is in keeping with the message on this subject within much of neoclassical economics, we seek to underscore that part of 'learning' that relates to an awareness of ideas.[3] In the tradition of Austrian economics, which has contributed virtually all that economics has to say on the subject of the entrepreneur's role in generating economic activity, the value of such awareness—or 'alertness' to opportunities—has long been recognized as being of fundamental importance to markets (Kirzner 1973).

Reid Hoffman, the billionaire entrepreneur involved in LinkedIn among several other things, articulates his views on alertness in an entrepreneur as 'permanent beta'. He describes it as a feeling that you must always be learning, and that you know things but don't know the whole game—you are alert to how the game is changing (Lewis 2018).

---

[3]Mainstream economics has given us a wealth of literature that emphasizes the role of learning in economic activity, and there has long been an emphasis on learning through a number of different channels that enhance human capital. Beyond efforts in the classroom, these channels also emphasize the merits of investing in research efforts, the benefits of colocating among others with access to knowledge, and gaining human capital through a process of 'learning by doing'. There has been commensurate attention to the vital role of firm- and industry-level effects as well as on macroeconomic frameworks that guide the incentives to learn. And while there has long been a realization that learning is intrinsically bounded by nature in rational individuals by dint of its attendant costs and extant technological limitations, behavioral approaches in the field have made it yet more clear how learning is also fundamentally affected by variations across people in their foresight, impatience, sense of fairness, and so forth.

A proclivity for awareness to variegated ideas very likely precedes entrepreneurial success; some research based on tracking the choices on courses made by alumni of the Stanford University's Graduate School of Business, for instance, showed that a varied curriculum was indicative of a later propensity for entrepreneurship (Lazear 2004).

However, while entrepreneurs may be especially dependent on this ability, in a theory of ideas it is more broadly relevant to the creation of all ideas. The process of rule generation in an idea space requires an individual to consciously and deliberately associate ideas within the idea space in the creation of a rule and, equally, to be receptive to information an individual receives from the external environment and then deliberately process the associated rule into its constituent ideas that are within the idea space.

Both of these processes require a degree of passive awareness *and* active awareness in the individual of the pertinent ideas; passive awareness can be seen as a general 'receptivity' to ideas and active awareness can be understood as purposive 'search' for ideas.

One could, for example, contrast the approach taken by Jackson Pollock to that taken by Steve Jobs. Pollock famously remarked about the process of his work that, when he was 'in his painting', he was not aware of what he was doing. In other words, he painted with a passive awareness, unafraid to make mistakes, and permitting his process of ideation to reflect freely in his works. Steve Jobs, on the other hand, described a much more actively aware process when he suggested that the key was to purposefully work toward achieving the goal of simple ideas by working hard to getting 'your thinking clean'. In other words, he was describing a process of active awareness, where the process of rule generation had a clear objective.

Jazz is another interesting case in point. It involves improvising extemporaneously over underlying chord progressions that can themselves be exceedingly complex. A soloist has to find the right mix of passive awareness and active awareness and often adjust it several times during the course of a single song. The structure of the whole piece has to be built conscientiously with an active awareness, yet the ideas necessary for a truly creative solo piece require an equally conscientious receptivity to ideas to associate with that structure. Charlie Parker, the jazz improvisation virtuoso, illustrates this two-pronged approach to ideation in jazz when he observed: 'Music is your own experience, your own thoughts, your wisdom. If you don't live it, it won't come out of your horn. They teach you there's a boundary line to music. But, man, there's no boundary line to art'.

Awareness of new ideas assists any process of developing new rules within the individual's idea space, whether it is by a deliberate association of ideas (both new and extant) to construct other new rules or by a deconstruction of new ideas into constituent ideas within the idea space.

More generally, note that both active awareness and passive awareness of new ideas have the potential to alter the centrality of any given idea both upward and downward within an individual's idea space, thereby altering the overall structure of the individual's idea space dynamically. Over time, an idea that is crucial to generating more rules is far more likely to become more central and possibly even become a member of the idea space's core. Once an idea does become a core rule, it also becomes more prominent within the worldview of an individual and more likely to bias an individual's awareness for those ideas that can then be associated with it.

2.2.2 Density

The production of rules from associating ideas entails a generic cost that essentially hinges on the simple notion that increasing conceptual complexity entails a larger cognitive effort (Flum 2003). Stated differently, awareness in the individual is a limited resource, and its use comes at an increasing cost to the individual in the generation of rules.

Even when we extend this premise to the realm of artificial intelligence, the relevance of a cost remains relevant. Neural networks are trained—which is to say, they learn—on the basis of a cost function that captures the error that an algorithm produces in its outputs as it tries to achieve some objective, for example, the accuracy with which a robot identifies obstacles or the frequency with which an image classifier disregards false positives as it scans a medical database.

It is in this context that the Allee effect is perhaps most useful in its insight—the generation of useful rules depends on the *density* of ideas within the idea space.

Sparseness of ideas within the overall idea space naturally cannot be expected to be conducive to any fecund process for rule generation, be it an incremental and sequential assembly of ideas or an experimental and combinatorial packaging of ideas, or some combination thereof. It is perhaps unfair then to see innovation as being characterized by Eureka moments to be a myth. In several idea spaces, it is possible to conceive that ideas might simply be exceedingly rare and, as a result, an inordinate amount of awareness must be actively applied to associating ideas in the construction of new rules. An entire career of a mathematician or a chef might be spent in finding a proof for one problem or in the creation of a single novel recipe. The effort may be so high that their efforts may justify being seen as outliers, or as truly Eureka moments for the individual and perhaps even for the community. Conversely, though, when far too many ideas begin to compete for the individual's limited awareness, the result is a diminished likelihood that better rules can successfully be developed.

Few great minds better illustrate the notion of trying to seek some optimal range for the density of ideas than Thomas Edison, who, rather famously, not only set explicit quotas for the number of ideas he wished to generate within a given space of time (one minor invention every ten days, and a major one every six!), but insisted that those workers around him should establish similar goals, as well. His intention was not to achieve success at that rapid pace, but to generate ideas regardless of the result. Failures were not much less valuable to him than the successes since they made the collective idea space of his laboratory dense with ideas.

Given our conceptualization of awareness and density, it becomes quite evident why specialization in one field of investigation (or, at the most, a few fields) is usually necessary to the development of exceptional insights based on complex ideas. Naturally, those individuals who have a comparative advantage in processing rules for a given idea subspace may not possess that same ability for all idea subspaces. Such individuals are more likely to be able to develop rules that are relatively significant only for a subspace of expertise.

There is, of course, a downside to this. History is littered with examples of specialists at the helm of an idea subspace who are unable to look beyond their worldviews for a seemingly obvious solution until someone 'thinks laterally' for a solution. Ignaz Semmelweis, instituting the simple practice of washing hands by doctors at the Vienna General Hospital in the mid-nineteenth century in order to prevent deaths among pregnant women from puerperal fever, is perhaps the most famous such example. Worldviews then simply could not permit germ theory as a viable mechanism for the spread of diseases. The phenomenon has come to be known as the Semmelweis reflex in medicine. However, it is widespread enough, incorporating all forms of idea subspaces imaginable, that it is also acknowledged as a deleterious form of groupthink.

2.2.3 Types of Idea-People

In sum then, the awareness of ideas and the density of ideas are defining characteristics of an idea space that work together, albeit with an essential tension that is basis for a Goldilocks solution for idea spaces. Awareness of ideas is essential to the creation of a dense idea space that holds greater potential to yield several new rules. However, awareness is also the chief mechanism to expand the overall size of an idea space, either through an active awareness directed toward specific new subspaces or through a passive awareness for new ideas, regardless of where they emerge.

As awareness expands the scope of an idea space, it naturally also reduces the overall density of ideas, lowering the overall potential of the idea space to generate new rules. There are, however, interesting combinations of the awareness and density of ideas in an idea space that influence the types of potentials that an individual may have with regard to creating new ideas.

In Fig. 2.1, we can see these potentialities in individual idea spaces visually by plotting awareness on one axis and density on another, and concentrating attention on four extreme combinations of awareness and density.

The light gray zones represent smaller individual idea spaces; when a low level of awareness is also characterized by a low density of ideas, the individual idea space is reflective of ignorance. With a higher density of ideas within one or two subspaces, the individual's overall idea space then becomes conducive to the application of specialized skills within that subspace. Such an individual may then be able to expand her idea space,

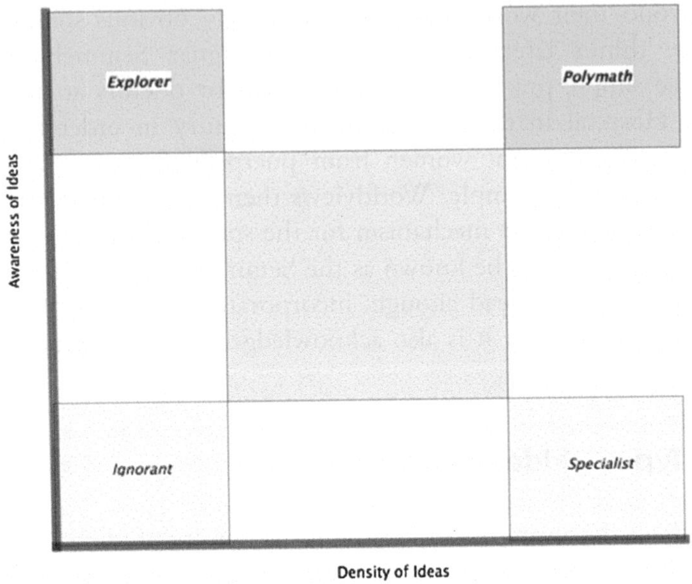

Fig. 2.1 The awareness and density of ideas in an idea space

as the darker gray zones depict; if a higher level of awareness is achieved while maintaining the same density of ideas across more subspaces than she had within the subspace of her initial skill, she would transform into a polymath. This is more likely to occur with active awareness expended across several subspaces in a somewhat more balanced manner. If, on the other hand, in expanding her idea space she was to lose the density of her of idea space, she would be more likely to become an explorer of ideas.

These are static and general characterizations, to be sure, but they are also quite useful to understanding the roles of awareness and density of ideas to our framework.

2.3 Subjectivity and Application: The Fuzzy-Crisp Continuum

Our preceding discussion also suggests that even identical levels of awareness and density for the same set of ideas do not necessarily result in identical effects on different individual idea spaces. This is a fact that is at the crux of our theory's conceptualization of subjectivity. In this regard, the manner in which we define all ideas as composites of other ideas is key to understanding how individuals may differ in their perspectives.

Bertrand Russell observed that words usually have a factual intention as well as the propensity for variegated emotional intentions. He suggested, as an example, that the intention in the idea of 'being firm' can quite easily be emotionally charged to different degrees in negative ways by describing it alternatively as 'being obstinate' or even as 'being pig-headed'. This observation, called Russell conjugation, quite starkly illustrates that subjectivity is relevant even in instances where we may think that we are dealing with facts alone.

More generally, in our case this point can be appreciated more directly by making a distinction between information received from external sources and the information an individual generates for the benefit of others. Since information is always received in the form of a rule from others and then deconstructed by the individual as a collection of ideas according to her own understanding, rules can only ever be seen as commonly shared across a group when their deconstruction into ideas does not permit much scope for variance. We can call such rules *crisp rules*, to denote that their intentions, as governed by the scope of composite ideas that inhere in them, are less susceptible to subjective interpretation.

The adjective 'crisp' has been selected advisedly, derived as it is from fuzzy set theory. Sets of objects usually denote discrete boundaries. The cutlery in my kitchen drawer is a discrete set comprising a dozen forks, knives, teaspoons, and tablespoons. The set of transportation options from home to work is similarly discrete, including cars, buses, trams, and the train. These sets do not permit room for debate or opinion. However, often situations arise where set membership itself is a probabilistic event. Whom you include in the set of your circle of friends might be far less clear. Best friends, for example, might have the pleasure of being definite shoo-ins. Their membership to the set is crisp, in that it has very high probability. Other friends may give you some reason for pause, as you consider their probability of membership to your esteemed set. Their membership is much 'fuzzier'. Contrary to the definite circles that we use to represent sets with, fuzzy sets ought to be visualized as clouds with boundaries that are progressively blurrier the farther away you move from the center, and where there is no distinct edge.

The subjective interpretation of rules—that is to say, the manner in which an individual deconstructs any given idea into its composite ideas in the idea space—is a defining feature of our theory. This emphasis derives directly from the observation that any given rule or idea can only be *fully* crisp and, consequently, free from subjective interpretation across a group of people, if the mapping of that rule into its constituent ideas is identical across all its members. In other words, fully crisp rules must be such that, when deconstructed, the component ideas are also all fully crisp members of the sets that define the rule. All other rules—the vast majority of ideas—are not fully crisp by virtue of having at least some constituent rules that are also not fully crisp, and must permit at least some subjective variations across individual idea spaces, based on the degree of their fuzziness.

The relation between subjectivity and ideas is hardly a novel premise. In management theory, for example, higher diversity across the members of a group, especially when it is mediated within an effective overall organizational process, is thought to be a strong determinant of the group's innovativeness (Enriqueta et al. 2005). It ought to be mentioned that it is the diversity of *ideas* that we are interested in; diversity of individuals is relevant to our discussion to the extent that it is strongly correlated with a diversity of ideas as well, which may, of course, not necessarily be the case, even if the two are frequently conflated. Observations of diversity enhancing an innovative mind-set are routinely confirmed in a variety of contexts, such as the social media habits of employees or their performance in the presence of an outsider (Janssen et al. 2004; Parise et al. 2015).

Generally, ideas in our theory follow an entropic principle, not too dissimilar from Shannon's entropy in information science: As new rules are recursively generated from component ideas derived from fuzzy sets, they tend toward higher overall fuzziness, and it is this that provides more complex rules the potential for a higher information content. In a more limited but obvious way, this has been observed in the context of classroom learning. The learners may be taught the same concepts and may even be examined on their understanding of the rules that govern those concepts to ascertain that they are equally familiar with them. However, when it comes to their application of the concepts in practice, their sound understanding of the concepts is no guarantee that the applications will be equally sound as well. The application of a learned rule relies on the overall structure of an individual's idea space and, thereby, the construction of additional new rules that are contextual by definition; this increases the probability that the rules become fuzzier.

Researchers in psychology have long argued that the ability an individual has to recognize essential features of a particular situation where knowledge can be applied is key to their ability to apply learning effectively (Gorrell 1993). For an observer of a game, for instance, acquiring some rudimentary knowledge of the rules of a game does not give complete insight on an expert player's thought process—that is to say, the player's 'cognitive model'. Acquiring information on a second set of deeper rules that the expert makes use of in using her knowledge to play the game in a variety of contexts, however, does improve understanding and increase appreciation for the observer. This is not so much 'learning by doing', but learning by understanding how it is done in practice (Arrow 1962; Vahabi 1997). As a matter of fact, much research in pedagogical contexts has shown that this distinction has real value for learners.

In the context of the idea space, this constitutes a key insight. Application requires an individual using a rule, no matter how crisp it might be, to deliberately associate it with other ideas in her idea space to order to construct a different rule that guides her behavior. Note that such rules of application can never be crisper than the initial rules used to construct them. Across a group of people sharing a crisp rule, the applications are bound to produce sufficient variance owing to subjective interpretations over which other ideas in an individual's idea space might appear to be more or less pertinent. As such, in contrast with crisp rules, we can call these types of rules *fuzzy rules*.

The relation between the crisp rules of a game (the fundamental 'concepts', as it were) and the fuzzy rules that someone proficient at the game employs in given states of the game's play can also be considered under the framework of memory and recall—specifically, the idea that information of a foundational and conceptual nature is stored in an individual's long-term memory (LTM), and that it is retained there with more fidelity, as it were. Research shows that humans, and indeed several animals as well, retain in their long-term memory stores the type information that provides them with an evolutionary advantage (Cowan 2008). In other words, information that enhances an individual's fitness for a game dominates her long-term memory. Short-term memory (STM), on the other hand, is a more limited, though more readily accessible store of information; STM, however, suffers from decay, in addition to having clear limits on the discrete quantities—known as 'chunks' by cognitive psychologists—of information it can manage (Miller 1956; Koch et al. 2006). As its name suggests, information in short-term memory is swiftly lost unless it is constantly rehearsed; its function is to cope with the more rapid processing of information that cannot be stored away in long-term memory with sufficient ease (Lewis-Smith 1975). Finally, working memory can be seen as a close ally or even a component of short-term memory that acts as a sort of workspace for current problems tackled with effort. In conjunction with working memory, short-term memory is therefore entirely conducive to the creation of fuzzy rules that rely on crisp rules and the effortful association of them with other ideas using an individual's awareness.

2.3.1 Labels as Intention Flags

Variations will inevitably arise across individuals in their placement of an idea along its fuzzy-crisp continuum. This simple fact suggests how individual subjectivity in the perception of an idea interacts with the recursive and interconnected nature of ideas as they are shared across a group of individuals. This is a central point to our approach and it is worth considering it more directly.

We have imagined an idea space—be it one that pertains to an individual or to some broader domain of knowledge—as an inexhaustible universe of ideas. The location of any given idea within this universe of the idea space is immediately indicative of at least three things. First, its location suggests the subspaces to which it belongs. Second, its connections suggest what role it plays in the development of other ideas. The fuzzy-crisp continuum introduces a third feature: The location of an idea within an idea space is different when it is *subjectively* perceived differently. The first two features are influenced by the recursive and associative nature by which ideas are generated; increasing the number of domains of knowledge, for example, directly permits an individual greater ability in the process of sorting any given idea with more accuracy. Thus, the manner in which different individuals become aware of the same idea within their own respective idea spaces depends not only on their own levels of awareness, but also on the combinations and relative importance of the subspaces within their overall idea spaces.

With regard to the third feature on the significance of an idea's location along the fuzzy-crisp continuum, it is useful to consider an example. Imagine a scenario where two friends decide to each bring a ball to the beach to play catch with the following day. One friend brings a tennis ball and the other decides to bring a baseball. If we imagine that they had a good deal of choice among alternatives, we can even assume that their selections reveal that they consider their choice of ball to be ideal for the purpose. When the friends meet the following morning, they may both even attest that the tennis ball and the baseball are equally good representations of what a 'ball suitable for playing catch at the beach' means to each one of them.

However, it is also possible to imagine that one friend brings an American football, while the other makes one with a spare notebook and a stash of rubber bands by crumpling the first page into a sphere and adding successive layers of pages over it held together by the rubber bands. Now both the balls that they bring to the beach may vary a great deal in their ability to qualify as good representations of the ball that the other had imagined as suitable for the purpose. It is possible that the balls may each be perceived as 'merely adequate' or it is possible that they may be seen as just as excellent as the tennis ball or baseball would have been. These are, it must be stressed, *subjective* assessments of each ball by each individual.

In Fig. 2.2, we depict a possible subjective representation across a range of balls for one of the individuals. The vertical axis of the figure is a measure of probability, 0–1, and the horizontal axis is a characteristic of the ball that is evaluated subjectively by the individual, perhaps its 'ease of catchability at the beach'. This characteristic serves as a *label* for

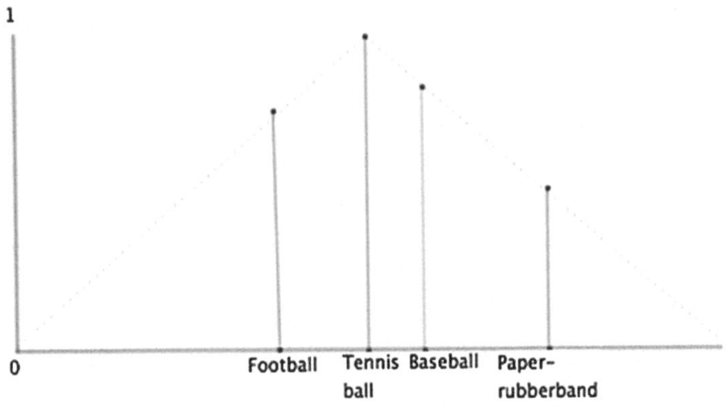

Fig. 2.2 Membership to the fuzzy set of balls for playing catch with at the beach for an individual

the idea pertaining to the set of balls for the purpose at hand; the label assigns the set some intentional purpose or objective associated with the idea, and each ball belongs to this set only with some probability.

The figure shows an example of a triangular membership function for objects to a fuzzy set—where the only fully crisp member would be found at the apex of the triangle, at a probability of one. Alternate shapes for the membership function may also be feasible, depending on the label for the idea, including trapezoidal functions and bell curves. The point to note is that, the lower this probability for any idea, the more tenuous its membership to this set. The idea of bringing the football may receive a lower probability score perhaps because it is seen as too hard to catch, whereas the paper rubber band ball may receive a low score for being seen as far too hard to launch. Neither is quite as ideal a member of this set for the individual in question as are the baseball and the tennis ball.

As the simple example above illustrates, subjectivity in a variety of situations can easily arise from an individual's model for associating a selection of fuzzy ideas within her own idea space; moreover, since each constituent idea is itself also a subjective assessment with its own label for a set of ideas with probabilistic members, the scope for subjectivity increases with the 'depth' of an idea. The fact that we routinely associate the word 'subjectivity' with an individual's ideas, even when they might be acknowledged as valid by others, reflects our apprehension of the possibility of imprecision represented by new ideas proposed by individuals.

To understand the nature of this imprecision, it is useful to imagine any idea or rule as a set of linked sets, or, more precisely, as a fuzzy set of linked fuzzy sets since each idea within a rule is a member of a fuzzy set, which is to say that it belongs to a set with a conceptual label that an individual assigns to it only with some probability. When this probability is low, the idea is a fuzzier member of that set and represents the concept less accurately. When the probability is greater, the idea becomes a crisper member of that set and captures the idea more accurately. A rule is, therefore, a selection of ideas associated by the individual to express a new idea, which is itself a member of a fuzzy set with a conceptual label. Thus, the imprecision in any rule emanates from the levels of fuzziness in its component ideas, as well as that of its own membership to the set it seeks to represent. The disciplinary mechanisms that an individual uses in building her subjective models for new rules are, of course, also types of rules themselves; even as metarules, mechanisms may themselves carry of degree of imprecision to begin with. There are, consequently, *several* potential sources for the imprecision in any given idea, and only the most egregious of these errors involve defining the objective for an idea incorrectly or using the wrong mechanism.

It seems rather appropriate to conclude this section with a quote from Einstein, who seemed to have presaged our views on labels and subjectivity in the formation of ideas when he remarked (Einstein 1995): 'The psychical entities which seem to serve as elements in thought are certain signs and more or less clear images which can be 'voluntarily' reproduced and combined'.

2.4 Fuzzy Rule Hierarchies and Clarification

An effective method to visualize the degree of fuzziness in an idea—seen as a composite of several other fuzzy ideas—is by building a *hierarchy of fuzzy rules*.

In such a hierarchy, the intention of the label of a fuzzy set is parsed more narrowly among constituent subsets; each subset covers a subjectively narrower range for variation in the intention of the idea than the set to which it immediately belongs. For instance, the friends in our example in the preceding section could have drastically reduced the probability that any one of them brings a football to the beach by specifying that the ball that they select must be 'spherical'—an idea that would have been far more likely to be crisp for both individuals.

The point is that the more granularity that is provided in the definition of a fuzzy set—by creating fuzzy subsets that span its range—the more likely it becomes that an idea that used to be a member with low probability will be more correctly classified within one of its subsets far more crisply. Doing this, however, also increases the complexity of ideas that rely on constituent ideas from lower levels within a deep hierarchy of fuzzy rules, by now requiring the rule to be based on greater specificity among the ideas that it contains. Consider saying 'bring a ball to the beach tomorrow' versus saying 'bring a ball to the beach tomorrow to play catch with, and make sure that it is spherical, soft, inexpensive and waterproof'.

Besides, rules with higher granularity do not necessarily result in lower fuzziness. This can be seen in Fig. 2.3, which represents the same idea on a set that has been split into three subsets and four subsets. With three, the original idea is now a member of each of the first two subsets with equal but lower probability than it was in the original set; with four, however, the second subset makes the idea a perfect representation of its intention. One can easily imagine several other constructions with different results.

The figure also adumbrates an important point in relation to innovation and creativity that is a central theme of this book: *New ideas often emerge as clarifications are attempted for the labels of ideas.*

An idea can be parsed into constituent ideas in more ways than one, and it is this process of enumerating the constituent ideas that we can think of as idea clarification. Note that clarification always results in a deeper hierarchy, with labels that carry a narrowing intention at each successive step. However, while the word clarification suggests that the process ought to result in something 'superior', it is important to stress

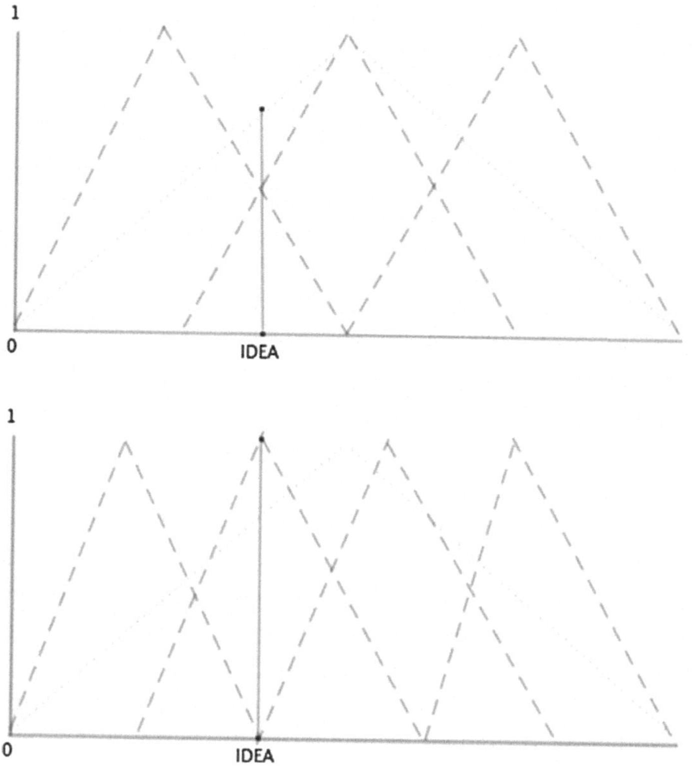

Fig. 2.3 A hierarchy of fuzzy rules with three and four subsets for the same idea

that, for a hierarchy of fuzzy rules, clarification may not necessarily yield a 'correct' or 'incorrect' outcome in a definitive sense. What matters is that clarification results in the development of further labels for constituent ideas, and the more of such labels for ideas that are developed, the greater the likelihood becomes that a new idea *may* be found that is a crisper member of the initial idea label.

This may also suggest that clarification should result in ever deeper hierarchies and, inexorably, lead toward crisper ideas—indeed, a set of fully crisp ideas in the limit. This, of course, is entirely possible for the case where the initial idea itself does not alter in relation to its associations with new ideas within a growing idea space. We shall see that this is rarely the case when ideas are contextually sensitive. Labels, in other words, are rarely static, and innovation seen as clarifications is, therefore, necessarily a continuous process.

A moment's reflection on the preceding should perhaps also suggest the immediate relevance of 'systems of representing ideas' to the creation of a deeper hierarchy of fuzzy rules—that is to say, a hierarchy with several levels of subsets that parses the intentions of each subset immediately above them even more finely. The ability to develop the depth of the hierarchy for a given fuzzy rule, with an increasing number of subsets that each represent greater specificity when compared to the larger set that the fuzzy rule represents, is constrained by a commensurate ability to express the associated ideas with suitable specificity. At least three observations are worth considering in this regard.

First, research on the preferences that individuals have for learning suggests that important variations exist in terms of how individuals prefer to engage with an idea. Some prefer learning through visual information, others prefer auditory channels, and yet others favor learning through experience; interesting differences also exist on the dimension of whether a learner prefers grappling with an idea in a practical and applied manner or in a more abstract and theoretical way (Kolb 2015). Yet, research in the field of neuroplasticity also suggests that these preferences are not set in stone, and can be altered depending on the frequency with which an individual faces a particular context for learning. Generally, therefore, we should expect that, especially when the modalities for learning are few and constrained, and when the context remains homogenous, the hierarchy for fuzzy rules should remain short.

Second, naturally languages play a key role in the ability of humans to represent and develop an idea. While languages can be of a pictorial and largely direction-free nature (i.e., semasiographic languages, such as cave drawings or computer icons) or of the more familiar written and directional nature (lexi-graphic languages), studies on their long-term evolution suggest that their overall structures (i.e., their syntax and vocabulary) evolve in punctuated bursts (Atkinson et al. 2008).

As the structure and usage of a protolanguage diverges for different subpopulations, one of the new languages can gather a new set of rules more rapidly than the other at first before returning to a more gradual pace of evolution. It has also been shown that languages spoken more broadly have simpler morphologies than those that are spoken more narrowly, but also have a greater phonological inventory (Nettle 2012). These observations suggest an interesting association between language and the hierarchy of fuzzy rules for ideas based on an impetus for individuals to innovate the system of idea representation when there is a more pressing need for the expression of more richness in their ideas.

Third, the development of the hierarchy for a fuzzy rule underscores the relationship between the cognitive demands associated in making a fuzzy idea crisper and the inherent ability of a 'language' to formulate and express the idea with sufficient specificity and accuracy. That this cognitive demand requires a sufficiently developed brain is often provided as a reason for why the emergence of a symbolic language was late in the evolution of hominins compared to the emergence of representational art (Rossano 2010). More abstractly, this is a point that also finds support among researchers in information science who study the analytical differences that exist between computationally complex problems and those that are descriptively complex in terms of the systems of logic that can be used to translate them (Flum 2003).

2.5 Collective Idea Spaces

We tend to see a collection of individuals that all identify with a group when the group itself carries an identity that serves some purpose. This purpose might simply be to identify some set of core ideas—codified knowledge, dogma, ideology, culture—that are crisp within the worldviews of all members of the group. However, if this was all that a group represented it would leave little room for any innovation at the level of the group, unless groups themselves came into contact with other groups with different worldviews; indeed, a useful way of understanding cultural evolution is exactly by examining how the boundaries of cultures evolve through their interaction with one another (Hartley and Potts 2014).

A somewhat different and useful way of looking at groups is by beginning with an individual idea space and explicitly addressing how it might differ from the idea space common to a group of individuals—which is to say, how we might consider a collection of individual idea spaces.

In order to do so, we can extend the idea of a commons to describe a situation where the resource at risk of being overused is ideas. The fact that it is hard to imagine 'overusing' ideas as a problem has not escaped the notice of social scientists; perhaps somewhat facetiously, the notion of a comedy of the commons has been proposed to describe the situation where the resource in a commons is knowledge. The comedic aspect emanates from the fact that several people making use of knowledge usually results in the stock of knowledge expanding rather than depleting.

However, as the previous section suggests, there is an aspect of idea spaces that should give us pause in the optimism for a commons that deals in ideas and maintains unfettered access. Free access essentially enables an easier pooling in of the ideas that individuals have and, as such, it is a proxy for higher density in our framework for individual idea spaces. A variety of outcomes are possible. The types of new ideas generated by the commons as a result of such pooling depend on whether the density of ideas increases as the label for the ideas that the commons represent remains fixed, or whether the size of the commons expands to incorporate more labels.

There is a wealth of literature in foundational economics that concerns itself with the public-good aspect of knowledge, and, from that perspective, the problems are decidedly of a severe nature. The fact that one can conveniently free ride on the efforts of others is especially problematic when the costs inherent in the creation of knowledge cannot be recovered by its producer. The rational reaction of the knowledge producer when faced with free riders, or the prospect of theft generally, is to either underproduce knowledge or, to the extent possible, seek to enter her knowledge into an alternate domain that does permit her to be compensated for its use. These are well-examined issues in the legal literature on patents and copyrights. On the other hand, a different aspect that is equally worth examining, though has not generated anywhere near as much attention, is that of the situation where individuals *willingly* share ideas and contribute their knowledge gained through effortful action. Since the premise in economics is to expect under-provision of public goods that suffer from opportunistic free riders, this form of selfless provision of knowledge is both interesting and counterintuitive.

A useful place to start is by looking at knowledge commons, which we shall revisit in the following chapter. Knowledge commons represent a range of open access arenas that specifically encourage the collaborative production of knowledge. What is interesting about them is that a fundamental feature of such commons is an environment of shared principles among a group of individuals, or what we may see as relatively crisp rules on the labels for ideas.

By their nature, crisp rules can be shared among a group of individuals without increasing the risk of their distortion, which, in turn, permits a relatively higher degree of willingness to be collaborative among individuals who are members of the knowledge commons than with those who are not. The hierarchies of fuzzy rules that apply to these shared crisp rules begin from a common premise, even as they are further developed and clarified by the individuals. The crispest of such rules become essential to the core rules of the overall knowledge commons. As such, they form a reliable basis for the assessment of the compatibility of individuals to the knowledge commons. With shared crisp rules, such knowledge commons can usefully be seen to have a shared idea space, or what we can call a *collective idea space*. We shall see that this idea of a collective idea space permits us to keep track of the role of the contextual environment of a range of social groups.

A key point to consider is that a collective idea space aggregates over its corresponding subspace represented within individual idea spaces; the process of this aggregation, no matter how perfect, is limited by the extent to which the ideas that are represented across subspace mechanisms are similar.

When ideas pertaining to the labels themselves are relatively crisp across the individual idea subspaces, the collective idea space is also more likely to represent them as relatively crisp ideas and maintain a narrower overall focus—mutatis mutandis for fuzzy ideas. The aggregation across individual idea spaces over a particular subspace relies directly on one or more subspace mechanisms; as metarules, subspace mechanisms routinely employ the core ideas that are essential to informing the worldviews of the collective idea space as well.

A defining feature of any collective idea space, then, is its set of core rules—which may include others besides those that are embodied within a subspace mechanism—that enable a cohesive worldview for its members. The more cohesive this worldview is (that is to say, when it is based on a very crisp set of core rules), the more amenable it becomes to aggregating across individual idea spaces with a lower probability of the collective idea space's worldview being altered through a growing membership. As an extreme example of this, consider an application of the Internet of Things, or IoT, based entirely on dumb sensors relaying data that are then analyzed by an AI algorithm; the system's very appeal is that it is theoretically scalable to any number of sensors desired, creating a collective idea space of a size that is unimaginable with humans.

However, such a collective space would be infeasible when the individual idea spaces are complex, since its appeal is reduced to those individuals who may only have the core ideas of the collective idea space as fuzzier members of their own individual idea spaces. This is an idea that we shall develop further in the next chapter, but the message should be clear upon some reflection even now. A broad and cohesive worldview that permits a large number of individuals to subscribe to the collective idea space, yet with only a low probability that any individual's idea space would threaten to alter its core rules, should be relatively rare. As a matter of fact, we shall see that such a collective idea space has a special status.

2.6 The Foundational Idea Space

What we ought to expect in a society is a nested set of collective idea spaces, each representing distinct mechanisms and generic worldviews. Some of these worldviews are compatible with others, some stand in opposition, and most are entirely unrelated.

Just as an individual idea space is capable of holding several subspaces that represent distinct mechanisms—indeed, often even some that represent mechanisms that are in opposition with one another—a society, too, has a similar structure of collective idea spaces. Regardless of this interplay of subspaces in an individual's idea space, the core rules that eventually represent her worldview cannot be held in opposition with one another when they do not redound to the benefit of the individual and to the cohesiveness of the individual's idea space.[4]

Similarly, at the broadest level in a society—that is, at the *highest* degree of aggregation—all idea spaces are members of an overarching collective idea space. Such a collective idea space has a rather unique set of core rules; to distinguish them from other core rules, we call them core principles. We shall examine these principles more carefully in Chapter 3. This overall collective idea space is the *foundational idea space*. All individual idea spaces and even other collective idea spaces at lower levels of aggregation are *all* members of the foundational idea space and cannot endure if they oppose its core rules.

All idea spaces in a society are derived from the core principles of the foundational idea space in the sense that they belong to the hierarchy of fuzzy rules where the core principles are seated as an anchor at the highest level. As the core principles begin to be defined more narrowly—according to varying contexts, different labels for ideas, and subspace mechanisms—the degree of aggregation across individual and collective idea spaces falls. Each knowledge commons—be it a community, a firm, an institution, and so on—represents a particular collective idea space with its own core rules that permit it to aggregate over different degrees of individual and collective idea spaces. However, much like a tree of life, *all* these core rules can be placed on a hierarchy of fuzzy rules atop of which are the core principles of the foundational idea space.

Within a collective idea space, core and crisp rules are a key basis for the development of fuzzy rules. Crisp rules are, by definition, deeper

[4]Cognitive dissonance, from the perspective of our theory at least, could be seen as a result of opposing core rules held simultaneously within the same idea space; Leon Festinger, who coined the term, suggested that a situation of cognitive dissonance demands a correction in terms of a change in behavior, the manner of cognition, an explicit suppression, or some form of rationalization.

down within a hierarchy of fuzzy rules and innovations are produced by combining them with other fuzzy rules in order to reduce their fuzziness. As such, the ability of a collective idea space to share crisp rules cleanly—which is to say, without increasing their fuzziness—is conducive to the production of innovations for the group. The inclusion of individuals in a knowledge commons, for example, by virtue of some shared crisp rules in a collective idea space permits a distribution of the responsibility of innovation across all 'members' of the group.

An essential question then becomes the desirability of an innovation produced by the collective idea space. Collective idea spaces, such as political institutions or religious groups, may not seek, or indeed even welcome, innovations of fuzzy rules on the basis of their shared core and crisp rules. In other words, while some collective idea spaces may seek to grow by deliberately innovating, as we have defined it, others may prefer preserving the structure of their collective idea space. We shall see how the core principles of the foundational idea space permit us to account for this variability.

Regardless of whether collective idea spaces have worldviews that accord with particular innovations, we can consider the broader question on how fuzzy rule innovations are shared—or, indeed, prevented from spreading, as the case may be—across everyone in the collective idea space. Why, for example, would a member of a collective idea space voluntarily share his or her discovery of a particular fuzzy rule with everyone else or, for that matter, voluntarily exert efforts to quell innovations of certain new fuzzy rules? To address all these challenges, the group needs to rely on a mechanism, much in the same way that mechanisms bind ideas within an individual's subspace.

For a collective idea space, one such key mechanism is that of reputation, or *prestige*, which is automatically ascribed by the social group to an individual undertaking particular actions that benefit the group. The designation of a 'Master Inventor' at IBM and the system of tenure at research universities are examples of prestige being provided to individuals for seeking actions from them in return that benefit the entire group; examples multiply rapidly when one examines the variety of public rewards given to individuals of all ages and groups of all kinds.

This is a key idea, and we will revisit it in the following chapters as well. To the extent that certain actions are more likely than others to generate social approval and reward the individual with prestige, she can be guided to undertake actions that may not directly benefit her in any other manner than permitting her to accumulate prestige. The fungible nature of prestige may perhaps justify the rationale for its accumulation by an individual generally, but, for the context of the collective idea space, it serves a specific purpose: *prestige weights ideas*, increasing their relative awareness for all members in the group.

Consider the collective idea space of a knowledge commons where two opposite conditions are both addressed by this mechanism of prestige. In the case where innovation is desirable, individuals with more prestige are able to asymmetrically influence the direction of the group's efforts on the discovery of fuzzy rules; in essence, they leverage their prestige to bias the exertions of awareness across the broader group toward the ideas that they favor. In a race for prestige, other members seek to out-innovate those with higher levels of prestige in order to generate a reputation for themselves, expanding the size of the collective idea space even further in their search for new ideas.

The acts of voluntarily sharing the innovation as visibly as possible are, of course, a necessary condition for the accumulation of prestige. Abandoning more efficient and optimal ways of doing things in favor of doing them more visibly serves as a costly signal to one's peers and redounds to one's prestige within the community.[5]

Similarly, the acts of rejecting an innovation is also a feature of the prestige mechanism in a collective idea space; rejected ideas are either those that compete against ideas proposed by individuals with more prestige or genuinely represent threats to the worldview of the collective idea space, or both. Under this framework, the rationales for secrecy or isolation are few, including perhaps a desire to wait till a fuzzy rule of higher prestige value can be expressed more crisply by the individual. In both baseball and cricket, pitchers and bowlers, for instance, have been

[5]The classic reference in economics is Spence (1973) in the context of education choices relevant to the job market. See McGuire and Hildebrandt (2005) in the context of hunting-foraging strategies of early human societies.

known to wait months, even years, to reveal a particular kind of delivery that they have been honing in private.

For the case where innovations are not expressly desired, and the group wishes to preserve the structure of the collective idea space, the group institutes a crisp rule for the distribution of prestige that favors centralization. Not only does this dull any desire by other members to innovate or share their innovations with others, it also ascribes the responsibility for any fundamental changes to the collective idea space to a select group who gain prestige from preserving the collective idea space and reducing the awareness of new ideas that may be used to develop fuzzy rules that are seen as undesirable. From a meme-theoretic evolutionary perspective, the group can be seen as acting as an altruistic entity in the preservation of the crisp rules, helping them, as it were, to endure the test of time.

2.7 Playing with Fuzzy Rules

It is perhaps worth reiterating the fact that the distinction between fuzzy rules and crisp rules is not a dichotomous one. It is more helpful to imagine ideas as being on a continuum of fuzziness to crispness, just as set membership is treated in fuzzy set theory. To see this vividly, sports are an especially illustrative place to look. While we will return to sports in Chapters 5 and 6 as a key application of our theory, it is worthwhile making a few observations at this juncture to motivate the argument.

One obvious purpose for the existence of a set of relatively crisp rules in any sport, in the guise of 'laws of the game', is to constrain the scope of behavior that its players are permitted to adopt; conformity with such rules enables the collective idea space of a sport to maintain a stable set of core rules that can form the basis of a worldview for its members, and then ensure its stability when the collective idea space is expanded to include more members and variegated contexts.

Yet, a fundamental component of a large variety of sports is also to *deliberately* leave a number of ideas within the collective idea space sufficiently fuzzy so that, so long as a player abides by the shared set of crisp rules, she can then use her insight and creativity in associating the crisp rules with those that have been left fuzzier. The ability to 'hit a ball anywhere in the park', 'throw a ball in whatever manner', 'move a piece wherever', 'use variations of whichever strategy', and so forth, are all examples of relatively fuzzy rules that are anchored to one or more core rules of the collective idea space of a sport by following each one with an injunction to do so 'as long as' some other core rule is not violated. Moored by the shared crisp rules, a game is, in effect, an idea space in action, every time it is played before us; participants employ their skill and creativity (or, perhaps, team strategy) competitively as they associate the various ideas in the game's collective idea space, sometimes developing new ideas and at other times making fuzzier ideas crisper.

To see how we might examine the process of individual creativity in sports, we can turn to a consideration of the subspace within an individual idea space that pertains to a particular sport. Perforce, those codified 'rules' of the sport that enable the individual to participate and develop further fuzzy rules that govern her own play must be part of the subspace core, reimagined as crisp rules. It can be argued that several essentially codified 'rules' need not even be explicitly codified if they are indubitably part of the core of the idea spaces of a sufficiently large number of individuals in the social group; withal, several other codified 'rules' that pertain to aspects that are unrelated to the individual's participation or play need not be represented as a rule in the core of her idea subspace. Indeed, the very existence of laws for games suggests that several of them may *not* pertain to the core of the idea subspaces of the players.

Using rules in the individual's subspace—both those that are crisp and those that are fuzzier—and her awareness of other pertinent ideas across her idea space, a player's creativity lies in her ability to generate context-dependent fuzzy rules that pertain to her own participation. The ability that an individual has to replicate such fuzzy rules, game after game, repeatedly, is idiosyncratic because only a fraction of the rules involved are crisp, and some may even be unrelated to the collective idea space of the sport. Some of these fuzzy rules may become crisper over time as the player develops her gameplay and learns the advantage that they provide her. Some such rules may even be introduced to the collective idea space of the sport and are then more likely to also become members of the subspace cores present in the idea spaces of other individuals.

To appreciate this point about context-dependent creativity in a different light, consider the theory of heuristics. Heuristics concerns itself with decision making in problem-solving situations by evaluating the nature of the shortcuts and rules of thumb that people rely on, and to what degree they can be effective in practice. Exceptionally effective heuristics are obviously beneficial in complex situations, and they are usually derived from a great deal of expertise and experience.[6]

In testing this theory with the game of chess, for instance, it was found that the simple heuristic of constraining the opponent into situations that permitted the least feasible replies was particularly useful among accomplished players (Simon and Chase 1973). The codified rules of the game of chess are 'common knowledge' to all players of chess, in that each player reliably knows that the rules are known to everyone, and that everyone knows this. In this sense, the rules of the game of chess are examples of crisp rules in the cores of the subspaces for the game in the individual idea spaces of every chess player. However, the heuristic of constraining players is not crisp to quite that extent.

Certainly, all accomplished players know this heuristic, but beginners may not. Besides, its execution requires knowledge of a range of fuzzy rules pertaining to chess, and even some that lie beyond, such as those that relate to the psychology of the opponent. The heuristic, therefore, is a very context-dependent fuzzy rule. A wide range of players, regardless of ability, is given an opportunity to innovate, in the sense that they can each use the heuristic to guide their behavior on the basis of how their opponent might be further constrained, as long as both players adhere to the core rules in their replies. The ability to play the unexpectedly good move, from the thousands of theoretically feasible moves, is contingent on the player's skill in perceiving the relevant search space for a given game's context effectively. Creativity and innovation are

[6]Interestingly, this capacity for accumulating heuristics in humans has even shown to provide us with an intrinsic advantage over computers in the context of playing computer games (Dubey et al. 2018).

contextual in the sense that the range of fuzzy rules that can be drawn upon in conjunction with an essential set of crisp rules is permitted to change in the game depending on its state of play.

The concept of a habit hierarchy in psychology also bolsters the idea of an immanent continuum of ideas in the behavioral responses of individuals (Hull 1930). A product of the stimulus-response framework in psychology, it suggests that a given stimulus can elicit a variety of responses from the same individual depending on the strength with which the stimulus correlates with multiple other stimuli that then work together on the individual's likely response. Since a context can vary in its presentation and emphasis of a stimulus, a given stimulus can generate a hierarchy of responses.

The relevance of this idea to our own framework is that it expressly considers the role that variability in contexts can have in eliciting changes in the behavior of an individual's responses based on perturbing some implicit habit or 'rule'. It suggests that a response is developed by an individual by allying stimuli of varying strengths; to the extent, we can see a stimulus as itself an operationalized idea and the strength of stimuli as being indicative of the fuzziness of an idea; there is a close parallel between the two approaches.

2.7.1 The Context of Baseball and Cricket

The quality of an accomplished player in baseball or cricket to effectively strike a ball within roughly half a second of its delivery (and much less in the case of table tennis) toward him depends in large part on an ability to not have to consciously think about all aspects of the chosen shot.

Research has shown that during the first phase of the delivery of a regular-length ball in cricket—the period prior to the ball bouncing off the pitch—the batsman compiles information on the flight of the ball and its trajectory, but only selects a shot in the roughly 200 ms remaining, after it has bounced (Land and McLeod 2000). This observation congenially accords with the heuristic provided to budding batsmen in cricket to try and play the ball 'as late as possible', and the compliment routinely leveled at the very best batsmen on how they seem to 'have a lot of time'.

This is a heuristic much like the one about constraining an opponent to fewer moves in chess, and its general vagueness (even to batsmen of some experience) suggests that it is seen as a fuzzy rule that is highly dependent on the context of the game and even the psyche of the player. This also suggests that the information gained from the first part of the ball's flight (which is very closely determined by the core rules of the game) is likely being used to retrieve information from the experienced batsman's long-term memory, which is then married with the information gained from the behavior of the ball much later in its trajectory, possibly even after its bounce (which is very largely context determined, and often quite random), from his short-term memory. The fact that it seems unlikely that all batsmen store information on ball trajectories between long- and short-term memory identically is perhaps further reason to suspect that the situation is better understood as one where creative individuals innovate by associating crisp and fuzzy ideas more effectively. A less accomplished batsman, by contrast, looks hurried and more exhausted because a great deal more of the information is worked out in short-term memory alone; perhaps more precisely, such batsmen have fewer crisp rules that they can rely on in their long-term memories, leaving their hierarchies of fuzzy rules far too broad and imprecise to deal with contexts that represent multiple variations.

In cricket, as opposed to baseball, the complexity of the ball's trajectory is compounded by the ball's bounce off the pitch, making the condition of the pitch a critical aspect of the game. Its general hardness, moisture content, cover of grass, degree of cracking and roughness, rate of deterioration over the course of play, and so on, are variables that are closely monitored by players, umpires, commentators, and keen spectators. At a variety of junctures throughout the game, batsmen can frequently be seen 'gardening'—that is to say, tending to small cracks and loose pebbles on the pitch where the ball might bounce, though excessive tampering with the pitch by any player is considered to be legitimate cause for censure. All of this underscores the fact that the batsman contends with a great deal more complexity, much of it highly contextual to the particular game, making it a lot harder for even the most experienced of batsmen to be able to develop entirely crisp rules for dealing with all types of ball deliveries.

In baseball, while there is no deviation of the ball off the ground to consider, owing to the configuration of the seam on a baseball, there is a greater variance of aerodynamic forces acting upon it in flight than for a cricket ball. Consequently, this idea of waiting tens of milliseconds longer to gather information prior to acting on a pitch has been shown to apply to baseball as well. Indeed, the function of the sight screen in cricket and the batter's eye screen in baseball are to assist the batter in gleaning as much information as possible from the initial flight of the ball—even the position of the bowler's or pitcher's arm and wrist at the instant just before the ball is delivered.

Research, based on the fMRI scans, shows that better athletes have a more developed ability to activate those particular regions of the brain that correspond to the task at hand. It also demonstrates that the overall level of neuronal activity in the brains of superior athletes is lower, especially in those regions associated with forming memories, with being overly conscious of one's motions and with emotional states. The effect of this finding on explaining observed characteristics of elite athlete behavior is interesting, to say the least (Sherwin et al. 2012). This pattern of brain activation suggests why superior athletes feel that they are 'in the zone' and are able to muster outstanding levels of focus: Associating crisp rules with fuzzy rules in context requires reducing the scope of ideas that the athlete needs to be aware of in the individual idea space. Awareness is a limited resource, after all, and its expenditure on the association of irrelevant rules detracts from the athlete's ability to play a better game.

With the region of the brain related to forming memories thwarted, it may also help explain why athletes often attribute their performance to supernatural causes. Superior athletes also have better developed connections between their musculoskeletal and nervous systems—partly a genetic advantage and partly developed by improving this connection through conscientious practice (Bascom 2012). It has even been shown that elite cricketers were better able to predict the trajectories of balls from a series of photographs when they stood with a bat in hand and imagined playing the stroke compared to when they were seated. Such 'forward models' of the brain are examples of the elaborate heuristics that athletes develop to predict outcomes that they wish to rely upon in the course of a game. These connections between their brains and their physical motions serve to improve their craft by helping them develop a hierarchy of fuzzy rules for their sport—one where more of the ideas relevant to any context are crisper.

References

Arrow, K. J. (1962). The Economic Implications of Learning by Doing. *Review of Economic Studies, 29,* 155–173.

Atkinson, Q., Meade, A., Venditti, C., Greenhill, S., & Pagel, M. (2008). Languages Evolve in Punctuational Bursts. *Science, 319*(5863), 588–588.

Bascom, N. (2012). Elite Athletes Get Their Heads in the Game. *Science News, 181*(1), 22–25.

Cowan, N. (2008). What Are the Differences Between Long-Term, Short-Term, and Working Memory? *Progress in Brain Research, 169,* 323–338.

Diamond, J. (2005). *Collapse: How Societies Choose to Fail or Survive.* London: Penguin Books.

Dubey, R., Agrawal, P., Pathak, D., Griffiths, T. L., & Efros, A. A. (2018). Investigating Human Priors for Playing Video Games. arXiv:1802.10217.

Einstein, A. (1995). *Ideas and Opinions.* New York: Broadway Books.

Enriqueta, A., Gilboa, I., Postlewaite, A., & Schmeidler, D. (2005). Fact-Free Learning. *The American Economic Review, 95*(5), 1355–1368.

Flum, J. (2003). Descriptive Complexity Theories. *Theoria: An International Journal for Theory, History and Foundations of Science, 18*(1), 47–58.

Golman, R., Hagmann, D., & Loewenstein, G. (2017). Information Avoidance. *Journal of Economic Literature, 55*(1), 96–135.

Gorrell, J. (1993). Cognitive Modeling and Implicit Rules: Effects on Problem-Solving Performance. *The American Journal of Psychology, 106*(1), 51–65.

Hannan, M. (2005). Ecologies of Organizations: Diversity and Identity. *The Journal of Economic Perspectives, 19*(1), 51–70.

Hartley, J., & Potts, J. (2014). *Cultural Science: A Natural History of Stories, Demes, Knowledge and Innovation.* Sydney: Bloomsbury.

Hauert, C., Holmes, M., & Doebeli, M. (2006). Evolutionary Games and Population Dynamics: Maintenance of Cooperation in Public Goods Games. *Proceedings: Biological Sciences, 273*(1600), 2565–2570.

Hull, C. L. (1930). Simple Trial-and-Error Learning: A Study in Psychological Theory. *Psychological Review, 37,* 241–256.

Janssen, O., Van de Vliert, E., & West, M. (2004). The Bright and Dark Sides of Individual and Group Innovation: A Special Issue Introduction. *Journal of Organizational Behavior, 25*(2), 129–145.

Jost, J. T., & Banaji, M. R. (1994). The Role of Stereotyping in System-Justification and the Production of False Consciousness. *British Journal of Social Psychology, 33*(1), 1–27.

Kahneman, D., & Tversky, A. (1984). Choices, Values, and Frames. *American Psychologist, 39*(4), 341–350.

Kirzner, I. M. (1973). *Competition and Entrepreneurship.* Chicago: University of Chicago Press.

Koch, I., Philipp, A., & Gade, M. (2006). Chunking in Task Sequences Modulates Task Inhibition. *Psychological Science, 17*(4), 346–350.

Kolb, D. (2015). *Experiential Learning: Experience as the Source of Learning and Development.* New York: Pearson Education.

Krebs, D. L. (2008). Morality: An Evolutionary Account. *Perspectives on Psychological Science, 3*(3), 149–172.

Land, M. F., & McLeod, P. (2000). From Eye Movements to Actions: How Batsmen Hit the Ball. *Nature Neuroscience, 12,* 1340–1345.

Lazear, E. (2004). Balanced Skills and Entrepreneurship. *American Economic Review, 94*(2), 208–211.

Lewis, D. (2018). *Entrepreneur Voices.* Irvine: Entrepreneur Press.

Lewis-Smith, M. (1975). Short-Term Memory as a Processing Shift. *The American Journal of Psychology, 88*(4), 605–626.

Liao, X., Rong, S., & Queller, D. C. (2015). Relatedness, Conflict, and the Evolution of Eusociality. *PLoS Biology, 13*(3): e1002098. https://doi.org/10.1371/journal.pbio.1002098.

McGuire, K. R., & Hildebrandt, W. R. (2005). Re-thinking Great Basin Foragers: Prestige Hunting and Costly Signaling During the Middle Archaic Period. *American Antiquity, 70*(4), 695–712.

Miller, G. A. (1956). The Magic Number Seven, Plus or Minus Two: Some Limits on Our Capacity for Processing Information. *The Psychological Review, 63,* 81–97.

Nairne, J., & Pandeirada, J. (2008). Adaptive Memory: Remembering with a Stone-Age Brain. *Current Directions in Psychological Science, 17*(4), 239–243.

Nelson, J., McKenzie, C., Cottrell, G., & Sejnowski, T. (2010). Experience Matters: Information Acquisition Optimizes Probability Gain. *Psychological Science, 21*(7), 960–969.

Nettle, D. (2012). Social Scale and Structural Complexity in Human Languages. *Philosophical Transactions: Biological Sciences, 367*(1597), 1829–1836.

Ostrom, E. (2010). Beyond Markets and States: Polycentric Governance of Complex Economic Systems. *The American Economic Review, 100,* 1–33.

Parise, S., Whelan, E., & Todd, S. (2015). How Twitter Users Can Generate Better Ideas. *MIT Sloan Management Review, 56*(4), 21.

Piaget, J. 1968. *Structuralism* (C. Maschler, Trans.). New York: Harper & Row.

Rossano, M. (2010). Making Friends, Making Tools, and Making Symbols. *Current Anthropology, 51*(S1), S89–S98.

Sachs, J., Mueller, U., Wilcox, T., & Bull, J. (2004). The Evolution of Cooperation. *The Quarterly Review of Biology, 79*(2), 135–160.

Schreiber, D., & Iacoboni, M. (2012). Huxtables on the Brain: An fMRI Study of Race and Norm Violation. *Political Psychology, 33*(3), 313–330.

Sherwin, J., Muraskin, J., & Sajda, P. (2012). You Can't Think and Hit at the Same Time: Neural Correlates of Baseball Pitch Classification. *Frontiers in Neuroscience, 6,* 177.

Simon, H., & Chase, W. (1973). Skill in Chess: Experiments with Chess-Playing Tasks and Computer Simulation of Skilled Performance Throw Light on Some Human Perceptual and Memory Processes. *American Scientist, 61*(4), 394–403.

Spence, M. (1973). Job Market Signaling. *Quarterly Journal of Economics, 90,* 225–243.

Stephens, P., Sutherland, W., & Freckleton, R. (1999). What Is the Allee Effect? *Oikos, 87*(1), 185–190.

Tversky, A., & Kahneman, D. (1981). The Framing of Decisions and the Psychology of Choice. *Science, 211*(4481), 453–458.

Vahabi, M. (1997). A Critical Survey of K. J. Arrow's Theory of Knowledge. *Papers in Political Economy, 29,* 35–65.

Watts, D. J., & Strogatz, S. H. (1998). Collective Dynamics of 'Small-World' Networks. *Nature, 393,* 440–442.

3

Mechanisms of the Theory of Ideas

The fundamental ambition of this book is to outline a practical framework for thinking about the pivotal role that ideas play for individuals and social groups, and there are five key components that serve as the building blocks for the overall theory. Before we consider each in turn, it is useful to understand the merits of the basis for our theory of ideas first.

While the theory has several moving parts, the essential message is simple: ideas seek long-livedness, and idea spaces determine the mechanisms. A multitude of factors can affect idea spaces, as we have seen, which is why understanding their characteristics provides insight into how ideas propagate and develop or stultify and retard.

While this may appear to be advocating for an anthropomorphic treatment of ideas, a key function that the theory of ideas serves is in suggesting why this assumption—that ideas ought to be seen as a fundamental unit of agency—is not misguided. It provides a common basis for considering ideation in humans across different contexts, as well as in intelligent systems of machines.

An individual is represented as an idea space in our theory of ideas, making her an embodiment of her ideas, on the one hand, but also making this embodiment less bound by her physical body, on the other;

P. Goorha and J. Potts, *Creativity and Innovation*,
https://doi.org/10.1007/978-3-319-94884-3_3

instead, an individual is conceptualized by the body of ideas that her idea space represents, which may be shared with other idea spaces in specific ways and holds the potential for continually being reshaped.

The possibilities that deep learning algorithms provide in building large-scale artificial intelligence systems make the relevance of this approach yet more vivid. Each contributing unit, or information-gathering node, in a system of artificial intelligent machines is an individual idea space that can generate tomes of contextual information that can be stored in repositories that constitute collective idea spaces. By being able to parametrically describe the equivalents in our theory for subjectivity, limits to awareness, and density of ideas, such systems operate in more programmable and tractable ways. This makes them seem innocuous. However, these features are also what makes them amenable to scalability to a degree that is inconceivable for humans. To the extent that the interaction effects across idea spaces can have genuinely insidious side effects, those who are chary about AI would seem to have a legitimate cause for concern.

While it is true that an idea does not 'fear' death or 'get excited' at the prospect of becoming famous, ascribing a general purposive nature to ideas enables us to explain their nature and understand some outcomes that transcend disciplinary boundaries. The elements of the theory that we have discussed in the previous chapter and those we summarize below are helpful in that regard.

As we have seen, at the outset, the individual would need to possess the requisite awareness of all the component ideas in order for the ideas to ever become associated. Even if the ideas all belong to the same subspace, the association would require that the individuals are able to use one or more of the extant subspace metarules, or mechanisms, effectively. When an appropriate mechanism does not exist, a potential new rule may first require the invention of a new mechanism. Innovating new ideas, therefore, places a degree of emphasis on an individual having access to core rules of a disciplinary mechanism as crisp ideas with her own subspace. To the extent that the ideas draw upon different subspaces, this association is further encumbered by the ability to use the right set of mechanisms.

The hierarchy of fuzzy rules that we introduced in the Chapter 2 indicates why it is that any new idea would be fuzzy so long as any of its constituent ideas are themselves fuzzy; to the extent that it is fuzzy, it risks being less amenable to successful transmission between idea spaces and less likely to form the basis for the development of other rules. That the precision or crispness of a rule enhances the ability of an idea to endure through time is apparent for this reason. A less fuzzy idea can be exchanged more readily. However, crispness too is not panacea. Crisper ideas form more reliable bases for the development of other rules, but crispness alone does not ensure that they will necessarily become integral to the development of a large number of new rules and, as a result, become part of the core in an increasing number of individual and collective idea spaces.

When ideas do generate a range of other new ideas, it is easy to see that they would become more embedded within an idea space and more likely to be integral to its associated worldview. However, even crisp rules that are part of a core are routinely supplanted by new mechanisms or even entirely new subspaces that often arise in broader collective idea spaces. The import of all this is to suggest that it should give us little wonder that few ideas ever achieve the status of being truly long-lived, let alone acquiring any mantle of immortality!

3.1 The Five Components

3.1.1 Complex Ideas and Perspectives

Let us begin our consideration of the five components of our theory with an observation: The fundamental feature of knowledge is its multidimensionality.

Before we can unpack this statement and examine its value, it is worth persisting with an equally terse restatement. Essentially, it involves a recognition of the fact that ideas can be accumulated with increasing returns in a path-dependent manner along multiple cognate paths. The point of path dependence is arguably the least contentious component of this observation. Be it in the context of an individual or a group of individuals, access to some set of basic ideas permits access to an even greater set of ideas. Simple knowledge precedes complex knowledge in a sense that a simpler idea can presage or intuit the possibility of a more complex idea, provided the simpler idea is suitably extended. That this process of the accumulation of knowledge is, in a sense, *dependent* on following a 'path' is a recognition of the fact that knowledge that is inherently complex cannot be accessed until all the simpler knowledge that is necessary to its apprehension has been accessed and assimilated first. This perforce introduces a degree of both *sequentiality* and cumulative *consolidation* in the process of the acquisition of knowledge from the association of ideas.

In the context of the idea space, this argument can be restated by making a few preliminary observations. First, simpler rules sequentially precede more complex rules, and, therefore, more complex rules will tend to involve more ideas than do the simpler rules. Therefore, complexity has a very definite meaning for the idea space: Fuzzy rules that can become crisper are complex.

Some theories in quantum mechanics, for instance, can be considered complex for precisely this reason, based on being initially very fuzzy while simultaneously also holding the potential, often realized, for being extremely crisp. As fuzzy rules become crisper, they become accepted, or consolidated, in the sense that they can be both deconstructed and reconstructed into and from their component ideas more reliably across idea spaces; crisper rules are less open to subjective interpretation.

The process—and indeed the prospect—of a rule progressing to losing complexity is not entirely straightforward. However, it is perhaps evident that a complex rule that results sequentially from a set of simple rules cannot be less fuzzy than the crispest of its constituent simple rules. Further, as the constituent rules themselves become crisper, the complex rule loses its fuzziness as well and can be used to develop other rules. Indeed, it may even become part of the core of an idea subspace. Therefore, the complexity of rules is a notion that is related intrinsically to the degree of their fuzziness, as determined by their constituent rules.

Note, however, that the reverse is not true. The existence of a fuzzy rule does not necessarily suggest that it is complex since it is entirely possible to generate a fuzzy rule that cannot be made crisper. A unicorn in this sense is a highly fuzzy rule that is not complex.

Second, this view emphasizes *context* in the evaluation of a rule's complexity. A rule can, for example, be either crisp or fuzzy depending only on the idea space of which it is a member. A complex rule will never emerge from an idea space where the individual or group is not aware of the constituent rules and may become crisper much quicker where the constituent rules are all already, or are in the process of becoming, relatively crisp.

Note that the preceding argument is a necessary part of the observation that this path-dependent process has increasing returns, which is, of course, the basis for a logic suggesting that one gets proportionally more return from a given amount of knowledge. This is arguably a less intuitive observation than that of path dependence. Path-dependent processes usually involve *decreasing* returns as rising costs and increasing complexity begin to reduce the feasible return from a given investment. Almost any industrial manufacturing process stands as a ready example that the output will at some point begin to decrease from a given amount of increase in the inputs. This is what delimits the size of even the most efficient and largest commercial enterprise one can think of. Similarly, even in the case of knowledge, it would seem that sequentially adding to knowledge would also face decreasing returns as the low-hanging fruits of easy knowledge have been exhausted. Why then should there be increasing returns?

There are two additional reasons, besides path dependence, that one can appeal to in seeing why ideas are materially different. First, ideas are 'infinitely expansible', a concept that seeks to convey more than simply that ideas are inexhaustible and can be shared without diminution; it is a somewhat more trenchant acknowledgment of the fact that ideas can also be built upon ad infinitum. Second, the notion of multiple cognate paths emanating from some given idea suggests that rules can also be generated from a *combinatorial* process in addition to a sequential one; a new rule can be generated from the admixture of known rules through a process of random trial and error.

It may be argued that even a combinatorial process for the generation of rules is never really random. That it is still guided by some crisp rules, perhaps in the guise of mechanisms, experience, or even solid intuition, and that these rules are provided, perhaps very indirectly, by the core rules at the heart of some subspace of the idea space, albeit to a lesser degree than is the case with overtly sequential generation of new ideas. An example of this approach can be seen in the process often adopted in the discovery of drugs within the pharmaceutical industry and, generally, in the discovery of new compounds with industrial applications (Danielson et al. 1997).

Thus, the final part of our observation brings attention to the fact that the ideas from one subspace in the overall idea space can sometimes be used to access ideas that belong to several other subspaces through a process of random selection of ideas in the generation of some new rule. Some ideas, in this manner, become useful in the generation of rules from a combinatorial approach, in addition to being part of a sequential process of rule generation.

However, in a theory of ideas we ought to be able to acknowledge *all* these essential features comfortably within the same framework. And we can do this rather simply by introducing the notion of a *perspective* to the idea space.

The number of other ideas that any given idea is directly connected with across the idea space can be seen as the degree of connectivity for that idea. Naturally, the higher this degree is for an idea, the more essential it is likely to be to the overall connectivity of ideas in an idea space. Moreover, it is also then more likely to be a crucial member of several other rules; it would feature in a variety of sequences of ideas, each of which can then be seen as a perspective on the idea.

An idea that has several perspectives naturally serves to increase the overall cohesiveness of ideas in the idea space. While this does not automatically suggest that the idea must also be crucial to one or more crisp rules, it does increase the likelihood of this being the case and makes it much more likely that it becomes a member of the core of the idea space as well.

Given this setup, we can come to grips with the multidimensionality of ideas more simply by envisaging each 'dimension' of an idea as arising from a different perspective on the same idea. Adopting different perspectives on a rule or an idea entails looking at different sequences that it involves.

3.1.2 From Perspectives to Aspects

It is worth considering that the perspectives over an idea can be relatively closely related to one another or wildly disparate.

The closeness of their relation to each other in an idea space is determined in large part by the subspaces to which ideas inhere that then yield the perspectives on the initial idea. However, this need not always be the case. Subspaces are, after all, subjective interpretations of components of the broader collective idea space that defines a subject or discipline. In the context of an individual idea space, subspaces are more fluid constructions; ideas within one can often be more closely related to, or more distinct from, ideas in another, depending on the core rules in the individual's idea space. Bearing this caveat in mind, to the extent that perspectives on a given idea are derived broadly from a single subspace, we can say that they collectively constitute an idea's *aspect*.

Since each subspace relies on at least one unique mechanism as the metarule or framework for its ideas, an idea's aspects suggest how it might be associated with other ideas using the mechanisms of one or more subspaces. For example, market prices provide one such powerful mechanism for the economic aspects of ideas; de jure rules provide a mechanism for legal aspects while de facto rules could serve as the basis of a mechanism for an idea's social aspect; key scientific principles provide mechanisms for the various scientific aspects of an idea, and so forth.

Thus, for a given individual's idea space—say that of a keen athlete's—an idea such as 'throwing a fastball' could, for example, have a biological aspect based in the mechanisms of evolution applied to the human arm, a physics aspect based in kinematics, a social aspect based on the mechanism of trust among members of a group, and an economic aspect based on the market value of the player being determined by the accuracy and power of his pitch. These aspects may then motivate the individual to develop his muscles and flexibility, learn the mechanics of the ball's trajectory, use his place within the team and relationship with the coach as a motivation for his efforts, and enable his agent to seek a higher reward for his skill in a league. For a different individual—perhaps someone without a serious interest in sports—the same idea may only have one aspect of relevance, say, for example, a social aspect that prompts him to play catch once a year with a group of friends.

Even this simple example makes it abundantly clear that aspects need not always relate to perspectives that belong to a single subspace alone. Consider those instances where a rule is generated combinatorially using ideas across different subspaces. In such cases, the new rule can then serve to expressly relate perspectives in different subspaces more closely than they may have been before. For those instances where a combinatorial rule is generated by linking ideas or rules across subspaces, the resulting rule can even generate an altogether new, hybrid subspace; when this occurs across a number of individual idea spaces, they can even be realized as collective idea spaces that represent new interdisciplinary approaches, such as econophysics, neuropsychology, or computational biology. The point is that, for an individual idea space, the discovery of a new aspect on an idea often holds the potential for a new subspace to be established. Thus, throwing a ball for a living may prompt an individual to develop a rule that combines a multitude of aspects into just two new aspects that inhere to the subspaces of, say sports management and sports psychology.

It is perhaps worth reflecting on how an individual might begin constructing a new aspect for an idea on an ad hoc basis, before we can examine the interesting problem of how such an aspect might become more broadly accepted and even feature in a collective idea space.

In psychology, a great deal of research has been done on understanding how an individual behaves when confronted with an excess of information to process in the execution of some task. In this context, some interesting research conducted in the 1950s proposed that, while processing information, individuals are only comfortable with tackling approximately seven discrete chunks of information at a time within their working memory Miller (1956). Further research since then has placed this at a somewhat lower number. While this observation is interesting as a test of the limits of working memory, what is especially noteworthy is the application of this 'chunking theory', as it has come to be known, to those situations where individuals attempt to switch between multiple tasks. In such situations, it has been observed that individuals tend to focus on those chunks of information that particularly help them transition from one task in the sequence to the next (Koch et al. 2006).

In other words, individuals develop big-picture schemata that enable them to relate the information they receive across the various tasks, and they generate chunks that specifically link the common information that associates the tasks. In our context of the theory of ideas, these chunks can be seen as the ad hoc aspects generated by an individual from perspectives over the associated ideas or rules that otherwise aren't immanent in an aspect within one of their subspaces. In this respect, mnemonics to memorize random information may be seen as an ad hoc aspect. The ad hoc aspect in a chunk constitutes a fuzzy rule relating ideas that are common across the perspectives over some set of ideas or rules that are presented to the individual. When some such fuzzy rules become crisper they then begin to hold the potential to yield subspaces.

3.1.3 Consolidation and Realizability

There are two forms of *consolidation* that are relevant to the theory of ideas: consolidation over an idea and consolidation over an idea's aspect, which can be seen as a second-order consolidation—that is to say, consolidation over ideas pertaining to an idea.

The former is the more straightforward of the two, suggesting a simple method by which ideas can become more generally accepted. Essentially, it involves new rules that yield perspectives that draw on extant aspects that are shared across the idea spaces of a sufficient number of people in a social group for the idea to become less fuzzy and, consequently, more broadly accepted. To see this quite simply, we can rely on the observation that a fuzzy rule developed within a collective idea space is more likely to become consolidated than one where a collective idea space does not exist between a group of people for that rule. If an idea is consolidated, it can said to have been accepted, in its current form, across a group of people in an identifiable social group.

A consolidated idea is also *realizable*, which is to say that the idea can be put into broad use. Using an idea can have a great many different meanings, of course, depending on the context. For our purpose, however, a realizable idea is simply one that permits relying on one or more of its known aspect mechanisms to achieve coordination across a group of people; the idea can then be associated with other ideas and used for the realization of other rules or embodied in its current form and realized within a fixed product or service.

What can we say about the number of individuals that might be needed before an idea can be seen as realizable?

The literature in sociology, epidemiology, economics, and marketing are all lively on the general topic of determining a critical threshold of individuals before an outcome switches from being unstable or transient to becoming stable and accepted; applications examined include, among much else, access to credit markets, entry into job markets, innovation diffusions, the spread of diseases, advertising campaigns, and religious adherence. A great deal hinges on structural features that characterize the various subgroups of people within a community—the characteristics of the distributions of one set of individuals that possess one particular trait influencing another set of individuals who lack that trait. Yet, while a focus on structural features (such as the characteristics of a social network or relative distributions of groups of consumers) seems intuitively accurate, what such studies usually also highlight is the equally important role of the mechanisms that the underlying 'idea' draws upon. A religious idea on prayer may require the mechanism of religiosity to enable its realization across an entire group of adherents; a political idea on a voting rule may appeal to the mechanism of patriotism and civic duty among the eligible electorate for its realization, and a sociological idea on social media participation might appeal to the mechanisms of both a market price for the platform it relies on and the level of social capital in the group to enable its wider circulation.

The second form of consolidation—that over an idea's aspect—is a central concept to the theory of ideas since it is a prerequisite to both the establishment of a new subspace from a new aspect for an idea and the broad acceptance of an idea across a group of people who hold different worldviews.

The conditions for a new aspect for an idea emerge when an individual's idea space permits her to either reorganize the perspectives of the idea or develop new perspectives on the idea, and to do this in such a manner that she resolves them into an aspect within which the idea becomes a crisper new rule than it was within any of the extant subspaces with accepted mechanisms.

While this proposes a new aspect for the idea, it does not suggest a new subspace, which would require a metarule to serve the role of a new mechanism. Consolidation for the new aspect incipiently emerges as other individuals search across their schemata of subspace mechanisms and discover that the new rule has the potential to serve as the mechanism for a new subspace within which other rules can be developed.

That this is a rather high standard for a new subspace to be consolidated should be obvious. The discovery of a new aspect for an existing idea, even when it does not resolve to one of the extant subspaces, is not sufficient. Only when this new aspect is also developed on the basis of a crisper rule does it enable the aspect to become consolidated, and only when this new rule also suggests a mechanism that assists in the development of other rules with similar aspects does a new subspace become more likely to be consolidated.

In the case of the natural sciences, this is perhaps clearest to see, and Thomas Kuhn's *The Structure of Scientific Revolutions* is inarguably the definitive study of the establishment of new subspaces, or what he calls paradigm shifts (Kuhn 2012). Each of the examples he considers, such as Darwin's Natural Selection and Einstein's General Relativity, all represented key departures from the established worldviews for the respective subject.

A new subspace is, however, not just relevant to the natural sciences. In the course of human history, new religions, political movements, social traditions, and market discoveries have frequently uprooted the established ideas, often very violently.

The consolidation of a new aspect emerges on the basis of a crisper and simpler rule being preferred by a group of people over fuzzier rules. The consolidation of a new subspace, however, suggests a marked change in the idea spaces of that group of people, as they use the mechanism of the new subspace in the development of new rules, often entirely abandoning the rules with aspects that inhere to other extant subspaces instead.

3.1.4 Bringing Context Back in

It is perhaps sufficiently clear by now that an emphasis on context is foundational to our theory of ideas. Context matters and it is the subjectivity that ideas embody that make it matter most directly.

Differences in context can arise from a number of sources, but they can eventually be understood as the differences in the subjective perception of an idea across individuals, or, perhaps more precisely, in the structure of the hierarchies of fuzzy rules across individual idea spaces. Consequently, in thinking about ideas using our framework we expect contextual differences to play a strong role in most scenarios.

Consider even the rather tame situation in which everyone in a group possesses the same set of crisp rules at the core of their idea space, giving them all very similar worldviews. If there are differences among the members of the group in their levels of awareness, this fact would already hold the potential to lead them apart as they come into contact with any new ideas. They would begin forming new rules sufficiently differently from one another; some of these rules would be shared across other members of the group with similar levels of awareness, permitting at least some fraction of the shared rules to become crisper. To the extent that these new crisp rules eventually become members of the cores of their idea spaces, it becomes entirely possible that their worldviews will begin to diverge from the rest in the initial group as well.

This difference may be slight; indeed, since the crisp rules in the core of a collective idea space are not frequently altered, we would expect them to be slight. However, with a sustained difference in the level of awareness or biases in the exposure to new sources of ideas, we are faced with a greater probability that a change to the core will make a palpable difference to worldviews. With nothing but somewhat different levels of awareness and a sufficient period of time, therefore, we can be led to a scenario where a homogenous group of individuals become fundamentally different in their outlook. If we had begun with a situation where differences had existed in the subspaces among the idea spaces of the members of this group as well, these departures in their worldviews would then have been even more pronounced.

Three pertinent issues deserve some further thought at this juncture. First, it is worth reiterating that, most often, several subspaces combine to provide the worldviews an idea space reflects, even when the idea space is a collective idea space that is grounded on the mechanisms of a particular field of enquiry, dogma, or philosophy. The generation of new ideas within any idea space perforce requires specialization within relatively few subspaces, and to use awareness in a more focused manner. In the case of cutting-edge innovation, one can readily imagine such specialization to even require concentration on a single mechanism within a single subspace.

That such a process of specialization is based on the sound principles of comparative advantage is hardly difficult to fathom. By its nature, an idea space operates on the principles of comparative advantage, namely specialization and exchange in an environment of some resource of limited availability; the association of ideas for the creation of new rules sequentially encourages specialization *within* subspaces, while the awareness of ideas in several subspaces in pursuit of new rules combinatorially encourages exchange of ideas *across* subspaces. Yet, comparative advantage also creates conditions that encourage contexts remaining sufficiently variegated across ideas spaces. Indeed, it ensures an increase in the relevance of contextual dissimilarities; fueled by an expanding supply of ideas, the need for specialization increases unboundedly, yet the level of awareness needed for this pursuit of new ideas always remains limited at any given time.

Second, we can specify the sources that may serve to prevent contexts diverging as essentially relying on some form of barrier, either to the awareness of new ideas or to the exchange of ideas across subspaces within the idea space of individuals and their social groups. We might consider the source of such barriers to be exogenous or endogenous with respect to the idea space. An exogenous barrier can be seen as one that arises from a collective idea space rather than an individual idea space. The banning of a source of ideas—an institution, a book or even an individual, for example—instituted at some higher level of social group insulates the individual's idea space from those ideas, thereby preventing them from possibly becoming a trigger for the context to diverge between the individual and her social group. Conversely, the barrier may be endogenous to the individual or collective, serving as a mechanism by which the idea space is deliberately delimited from ideas that compete for awareness, as might be the case with the indoctrination of a group of people to a particular set of ideas at the expense of others.

Nevertheless, the only method for entirely insulating an idea space—for making it entirely immune to context—would be to sequester the rules that the idea space contains from any form of association with other ideas and rules. The barriers would need to be entirely impenetrable! The form that this sequestration takes depends, to a large extent, on the degree of fuzziness of the rules themselves. This so because the perspectives, aspects, and mechanisms of fuzzy rules are not as well-defined as they are for crisp rules, and it is, consequently, harder to ensure that a fuzzier rule is not associated with some other idea. For crisper rules, with well-understood aspects, such sequestration is easier on account of the fact that, when shared among a group of people, it is done so with a much smaller chance of having been altered.

Third, a key source of variance in context arises from the coordinating role that mechanisms play for the realization of a consolidated idea. For a given idea with several aspects, some mechanisms can be relatively more or less important than others, depending on their efficacy in facilitating coordination. Moreover, across two social groups faced with an identical consolidated idea, we can certainly imagine the scenario that the groups do not draw upon the same set of the idea's aspect mechanisms to achieve its realization. And to the extent that a realized idea's development by the group is influenced by the aspect mechanisms that were used to achieve its instantiation, this initial choice over aspects makes context especially salient to its future. Once realized in a particular manner, the idea is then more likely to be developed using other ideas within the particular subspaces that provided the mechanisms for its realization, driving the wedge further between the contexts.

3.1.5 Recursiveness

A final key element of our theory for ideas is that of *recursiveness*.

To see why the process we have described is recursive, it is worth observing that the process by which new ideas emerge is self-similar, which is to say that it follows an identical mechanism even as the scale of the process is changed. Even at the most elementary level, it is the mere repetition of the process for the generation of new rules through the association of ideas in idea spaces that suggests recursiveness; every existing idea can be imagined as a construction of other ideas. In other words, we might deconstruct any initial idea and it would display exactly the same general properties that governed its assembly as would any new idea that then uses this initial idea.

It is also worth noting that, apart from their distinguishing attribute of infinite expansibility, ideas are also unique in that they operate along multiple orders of magnitude in time. Some crisp rules, even those that may once have belonged to the core of a sizable collective idea space for a substantial period of time, can become increasingly fuzzier over a time scale that extends over centuries. Equally, for a different collective idea space, one can imagine fuzzy ideas becoming rapidly crisper far more frequently, and over considerably shorter time periods as well.

Essentially, as more perspectives on an idea are discovered through time, different aspects for it also become more likely to emerge than had existed before, creating a new context that features substantively different mechanisms, or even new subspaces, than when the idea was originally formed. This often enables the idea to become consolidated and realized in meaningfully different ways than it had been before. Yet, regardless of the timescale, the process remains essentially and inexorably unchanged.

3.2 Theory and Praxis: Creativity Redux

Creativity is arguably one of the most researched topics in all of social science. Creativity research has a rich history, especially in the fields of philosophy and psychology. However, since 'being creative' is increasingly being identified as a highly desirable skill, it is receiving increasing attention from a range of other thinkers as well; it is seen as a vital attribute of the future worker, one that should be inculcated in students and built into their curricula and one that companies should seek to discover and enhance in its employees.

Yet, creativity and innovation are topics that, at their root, concern themselves with the processes of forming newer and better ideas. And so, it is perhaps no surprise that creativity plays an integral role in our theory of ideas as well.

Besides, *how* we can think about ideas in an effort to understand creativity, even why this emphasis on creativity makes sense in a world of markets with technologies that enable broadening integration is made vivid by our approach. Technologies are developed in order to enable the technical conditions that enable some practical advantage, such as capturing a larger share within a market or making life easier in some respect. Seen as such, technological improvements are rightly seen as innovations. We can understand the function of a technology to that of making marginal changes to a subspace mechanism. If a technological improvement were to result in a new *realizable* idea for a firm that then consolidates ideas across an even wider set of consumers, its market share increases. However, technologies also have the potential to improve a mechanism in such a manner that it yields new perspectives in extant ideas relating to other subspaces, thereby helping consolidate aspects across subspaces.

These integrative technologies are essentially based upon mechanisms that help consolidate aspects, and, as such, it often appears that they provide a measurable increase in observable collaborative activity across individuals. The reason for this is that they all feature essential core ideas that are more closely related to the foundational idea space for the broader social group. As we have seen before, the overarching mechanism that spans a foundational idea space is the broadest version of a technology; indeed, it has distinct core rules that we shall examine more fully in the next chapter. Any technological advance or improvement to a mechanism that brings it closer toward those foundational core rules would necessarily generate the basis for a broader collective idea space. Technologies that undergird the IoT, big data analytics, AI, cloud computing, gene editing, blockchains, and platforms that enable two-sided markets, all emphasize such consolidation of aspects, and, consequently, appear to suggest how markets can be integrated, approaching, in theory, a 'market singularity'.

We shall see that what is meant by creativity can be seen rather intuitively as a combination of two distinct bases of our theory of ideas. One class of creative activity lies in the skill of an individual for the identification and development of new perspectives for an extant idea or in proposing aspects for ideas that leverage extant mechanisms in order to make those ideas more realizable. A second class of the creative process can be found in proposing new mechanisms and new ideas.

There is of course 'real' creativity in both approaches, but the emphasis that each requires is different.

The development of new perspectives or the proposal of new aspects for an idea emphasizes the collective idea space more than the individual idea space, since any consideration of consolidation, let alone realizability, would require that the types of ideas within a collective idea space, and even its overall structure, be expressly considered. In contrast, the development of new ideas and mechanisms places a greater emphasis on the individual idea space. Here, ideas are associated without direct regard for their potential for consolidating the idea across a group, so long as they permit an idea to be associated within the individual's own idea space.

It is worth considering how these two bases for creativity differ also in their emphasis in terms of the precision of rules along the fuzzy to crisp continuum. Developing new ideas need not entail taking fuzzier ideas and, through association, making them any crisper; when consolidation is not an express consideration, this is especially the case. Indeed, even a new proposed mechanism for a subspace can similarly be less crisp than its constituent mechanism or mechanisms. However, the proposal of new perspectives and aspects for extant ideas in order to make them more realizable does place a greater emphasis on making the ideas crisper.

Another point worth reflecting on is the manner in which both these broad classes of creativity relate to the concepts of awareness and density that our theory relies on. Again, while both are obvious features in both forms of creativity, when the emphasis is on consolidating an idea within a collective idea space, it becomes a necessary part of the creative process for the individual's idea space to develop an awareness for the structure of ideas within the collective idea space. Merely by placing herself within a social group, the individual benefits from a passive awareness of ideas that can inform her creative process. Awareness is also a necessary feature of creativity based on developing new rules and mechanisms; however, here it is more directly used by the individual in associating ideas. Since consolidation is, by definition, a group-level phenomenon, the density of ideas within the group directly informs the individual in her creative process on the perspectives and aspects on an idea that exist and those that can be pursued. Conversely, when the objective of the creative process is new rules and mechanisms, the density of ideas is often of far less significance; generally, awareness is spent more liberally on associating ideas within a subspace rather than on increasing the number of ideas in the overall idea space.

Einstein famously observed that 'logic will get you from A to Z; imagination will get you everywhere'. Clearly, he saw two distinct forms of creativity, one structured and linear, while the other more free-form and with the potential to yield surprising outcomes.

The observations made above are the basis of our theory's perspective on two broad classes of creativity. When the emphasis is on the generation of a new rule or new metarule rather than the feasibility of a solution, the creative process is freed from the contextual mooring provided by the collective idea space. Individuals following this class of creativity have been described by social scientists as 'big C' creatives (Csikszentmihalyi 1996). These are the inventors that one thinks of immediately when thinking of creative geniuses. The genius appellation is, of course, applied ex post, but the objective of such individuals—and they usually do work alone or in very small groups—is always to develop theories with little regard to the wider context beyond a narrower focus within their discipline.

On the other hand, when new perspectives and aspects must be created for an idea, the emphasis is squarely on the context and the ability to 'think out of box'. This is commonly known as the process of divergent thinking in creative individuals in psychology, and it emphasizes precisely this type of 'free' thinking. In resolving extant and new perspectives into aspects with extant or new mechanisms, this second basis for creativity in our theory enables the idea to become crisp for a group of individuals and serves to increase the likelihood of its consolidation and realization.

In its focus on market-driven innovations, economics and management theory has routinely addressed this second form of creativity. However, even within these disciplines—and especially in the popular literature that comes from management theorists—there appears to be an inchoate realization that creativity often combines both these dimensions. Generally, however, since these processes are often undertaken together in an overall recursive process of the pursuit of ideas, they are conflated. Creativity is then seen as a single idea, whereas from the perspective of our theory it is useful to consider it distinctly as at least of two varieties, with rather different attendant processes and proximate motivations.

Using our setup, we end this chapter by distinguishing between seeing creativity, generally, as being theoretically inclined or focused on praxis. Indeed, it is principally this distinction that sits at the heart of our focus on the existence of two distinct types of creativity; we can call them *theoretical creativity* and *praxis creativity*.

An idea space is a representation of the cognitive limits of the individual within which subjective representations of ideas can be developed as new rules. The greatest creative outcomes of the most brilliant among all individuals are, as a consequence, still constricted by these limits. This is perhaps what makes the true genius among the theoretically creative exceedingly rare. When the emphasis shifts to a discovery of the extent to which idea spaces can be made to cohere across one another in an effort to realize an idea, the creative enterprise then has to concern itself with rules that enable such integration through the use of appropriate mechanisms. This form of praxis creativity requires a collaborative effort, if not in directly consolidating and realizing an idea then, at the very least, in helping define its relevant context.

References

Csikszentmihalyi, M. (1996). *Creativity: Flow and the Psychology of Discovery and Invention*. New York: HarperCollins.

Danielson, E., Golden, J. H., McFarland, E. W., Reaves, C. M., Weinberg, W. H., & Wu, X. D. (1997). A Combinatorial Approach to the Discovery and Optimization of Luminescent Materials. *Nature, 389*, 944–948.

Koch, I., Philipp, A., & Gade, M. (2006). Chunking in Task Sequences Modulates Task Inhibition. *Psychological Science, 17*(4), 346–350.

Kuhn, T. S. (2012). *The Structure of Scientific Revolutions* (50th anniversary). Chicago: University of Chicago Press.

Miller, G. A. (1956). The Magic Number Seven, Plus or Minus Two: Some Limits on Our Capacity for Processing Information. *The Psychological Review, 63*(81), 97.

4

Core Rules, Contracts, and Commons

4.1 Rules, Norms, and Transaction Costs

Much has been written by sociologists and institutional economists over the years on the role of formal rules and norms, or 'informal rules', in a wide variety of societies. Several interesting questions have been tackled, yet there are two focal points of debate. The first pertains to the question of how rules come to be reified in the social and political institutions of a society, and the second is on understanding the mechanisms for how any set of rules, once instantiated in enduring social and political institutions, enables economies to endure and develop.

Especially since the 1950s, a tradition of research has begun linking political and economic development, beginning in sociology and political science and then taken up by economics.[1] Increasingly sophisticated arguments center on trying to sort out what is perhaps the grandest of all chicken-and-egg problems—that of endogeneity between political institutions and economic development. Is it the case that economic development creates the potential for political development or is it to be

[1]Representatives of this vast literature are Lipset (1959) at the beginning and, more contemporaneously, Acemoglu and Robinson (2013). North (1994) is a readable broad assessment.

P. Goorha and J. Potts, *Creativity and Innovation*,
https://doi.org/10.1007/978-3-319-94884-3_4

believed that the development of political institutions creates the potential for economies to thrive?

It appears that this important question has the distinct flavor of evolution by natural selection. Broadly speaking, in the context of evolution, the underlying mechanisms of physics provide overarching rules for biological interaction, which then describe the basis for biological complexity in evolutionary time. In the case of a society's development, when its own rules are seen as the overarching mechanism, there is no other mechanism available to suggest what guides the process that enables *their* sophistication through time in being able to create complex outcomes. To put it rather simply, logic mandates that complex rules must come from putting together several simpler rules. This is a key premise for the hierarchy of fuzzy rules.

However, it also suggests that this association of rules be achieved on the basis of some stable and predictable process. It is here that the merits of the transaction cost framework—possibly the progenitor of the broadest sweep of insights in economics—seem to become most obviously useful. The message it has been used for to convey in the context of development is that long-term successes depend in large measure on the degree to which a society's institutions make choices that are specifically designed to reduce the level of and variance in transaction costs across all forms of exchange. This quest, essentially for low and stable costs of doing business and forming social relations, guides the process of institutional 'evolution' over time. Bad rules are punished by rising transaction costs that inspire only feeble outcomes; such rules necessitate reform, and obstacles to achieving such change magnify the damage to a society's chances over time.

In our theory, this approach retains its analytic leverage, though the class of transaction costs most crucial to it is specifically those that interpose themselves in the aggregation of idea spaces and the awareness of ideas.

An exchange—be it a market exchange, a social exchange or a political exchange—can be seen as an instance of individual idea spaces being aggregated into a collective idea space for a particular context. The analytic value in collecting all forms of exchange under the same roof provided by our theory is that we are forced to see the connections across different forms of exchange in characterizing what we mean by a 'transaction cost'.

A noteworthy related tension is in sorting out the essential role of norms in a society at any given moment. Some suggest that norms are especially useful in ameliorating those problems that are attributable to situations where individual rational interest runs contrary to the group and so prevents cooperative outcomes from being sustainable. Others contend that norms actually serve to reduce the unpredictability of behavior in groups, but that they do not expressly make cooperation the status quo (Elster 1988).

This distinction is a useful one in our context. Recall that crisper rules are more likely to become the predominant basis for new ideas, as well as those ideas that are shared between individuals. As a result, their misapprehension, misrepresentation, or outright denial weakens or breaks a rule that is intrinsic to the idea subspace, as well as the connections across idea spaces; violating a crisp rule jeopardizes the integrity of both the individual idea space and the collective idea space.

Therefore, a generally accepted norm in a society—not murdering your neighbor for his belongings, for example—is a crisp rule that reduces transaction costs in the idea space—even if they might increase other forms of costs—by virtue of their being stable bases for building a collective idea space. Even if I never intend on doing business with my neighbor, voting for the same party he supports, or going bowling with him, the mere fact that we abide by some minimum set of crisp rules permits us to become part of the largest social group that we both belong to, and where everyone accepts, without misapprehension, the crisp rule pertaining to not murdering a neighbor for his or her belongings.

The preceding example is rather extreme, but the fact remains that while not murdering your neighbor may intuitively sound like a rather obvious rule to most of us now, it does not appear to be crisp enough to recommend itself as the basis for forming a social group to which an individual can reliably consider herself a member. And as such, it cannot reliably be considered as a core rule for a collective idea space. To begin with, it does not cover maiming the neighbor (Would that need a separate and slightly less crisp rule?) nor does it protect people who are living farther down the street. In those respects, and several others, it still remains a very fuzzy rule indeed, and cannot be relied upon as amenable to being shared across a set of people, let alone as a member of the core of their collective ideas space.

The fact that religious doctrine and legal codes prohibit murder recommends them as collective idea spaces that have instantiated it as a core rule. However, a significant degree of variance has been established to exist across religious and legal codes in terms of the severity of sanctions they place on murderers and on the severity of the punishments they prescribe for the guilty (Baumer and Martin 2013). More specifically, this variegated treatment of murder across societies, and often over time in the same society, is said to be based directly on the larger social environment. This fact alone suggests that, at least across social contexts, even perspectives on the idea of murder cannot be seen as the basis for a crisp enough rule.

This then begs the question on what we can say about the broadest set of crisp rules that does permit us to usefully define a society as a social group. What set of rules are most likely to become core rules for the collective idea space and then shape the worldviews that the group subscribes to?

4.1.1 Evolutionary Thinking

While these are fairly broad existential questions, we can begin with some useful principles to guide us. And, in this regard, the evolutionary logic for cooperation is an exceedingly useful place to start.

That the idea of evolution by natural selection would, on its own merit, be classed as a crisp rule—indeed, a mechanism—is hardly a contentious claim. However, its relevance as a core rule for a society's collective idea space would (indeed, should!), just as naturally, also be seen as unsatisfactory. It is, however, the mechanisms that it suggests for producing ostensibly cooperative outcomes in groups on the basis of individually rational choices that provide exceptionally useful insight.

There have indeed been scores of efforts to clarify the theoretical foundations of evolutionary logic for sociological applications through a class of games that mimic the complexities of the natural selection dynamic that routinely produces cooperative outcomes.[2] Very generally,

[2]For an especially interesting example, see Nowak (2006).

it is the methodology for how the resulting prize from a cooperative enterprise is allocated across the participants that tends to be at the heart of the consideration. Several factors have been carefully examined in this regard, including the role of direct and subtle forms of communication and the signaling of intent; repeated interactions with an array of 'trigger strategies' that act as carrots or sticks; the particular composition of individuals that are included in or excluded from coalitions; and the type of common knowledge shared across participants.

While all this research has provided exceedingly useful insight on cooperative games, some of the swath of concepts that speak to the evolution of cooperation, however, do not inspire much confidence as reliable foundations for our theory on ideas. The reason for this is simple. Like the institutional theories mentioned before, they too seem to require some external mechanism to motivate their relevance for social groups. To see this, a few examples prove instructive.

Consider the idea of indirect and network reciprocity, which appears to entail humans exerting a great deal of cognitive effort simply in order to keep track of the evolving network of interactions so that they may collectively be in a position to reward prosocial behavior across a social group. Addressing the problem of how such calculations are made relies on understanding the channels for the communication of information. To at least some degree, this is assuming the answer if the channels for communication of reputation themselves arose because of some mechanism that encouraged prosocial behavior.

Alternatively, consider the idea of group selection, which attempts to resolve the fundamental problem of why individual-level rationality indicates selfish behavior, yet a wide range of successful groups feature apparently willing cooperative individuals. The answer often proposed is that since cooperative groups grow faster and create more 'offspring', cooperative groups tend to outcompete uncooperative groups, and the population generally comes to be dominated by cooperative groups on the whole. Here, the problem is that there seems to be little motivation for why cooperative groups necessarily create more duplicate groups apart from some exogenously imposed limit on group size. A resource constraint of some kind shifts some of the focus away from the group and to the particular characteristics of the resource itself. If so, the correlations do not seem to be quite as intuitively obvious; inexhaustible resources ought to generate competition, and exhaustible ones ought to generate cooperation. Some argue that there is some compelling evidence of precisely this, while others are unconvinced. While it is certainly true that the cooperative outcomes might be easier to motivate in smaller groups, we are again assuming that cooperation in the face of intergroup competition is a strong enough motivator that optimizing on cooperation would dictate group sizes in a variety of contexts.

It is worth underscoring the fact that the bulk of these objections are not concerns in the least among evolutionary biologists; with sexual selection, the mechanism is quite clearly understood as the longevity of genes, enabled through the information-carrying prowess that DNA represents. Scaled up to the level of individuals and societies, however, we are often forewarned by these groups of scientists that the same premise cannot be reliably applied. To wit, we need a similar and equally reliable mechanism for our context, and indeed a framework on ideas in social groups.

4.2 Foundational Core Principles

Based on the observations made in the preceding section, we can now deal with some aspects on the nature of the collective idea spaces for social groups that play a pivotal role for our theory on ideas.

To begin with, we might note that if cooperation is built on an evolutionary dynamic, it requires a mechanism that stands on its own, rather than one that is indicated by particular contexts and inapplicable in others.

In what follows, we shall develop the proposition that this invariant mechanism can be seen as being guided by the overarching four core principles of *safety*, *longevity*, *fairness*, and *control*. Further, the fact that these principles are inextricably intertwined with one another in observed outcomes makes it harder to separate them, but this synchroneity also goes some way in suggesting their special relevance as foundational ideas, or *core principles*, for social groups taken at their highest level of aggregation.

The highest level of aggregation is one for which the collective idea space is broad enough to cater to the scenario where all of its constituent individual idea spaces need not represent any commonality in their fuzzier rules as long as they all share in common these core principles. If, in a social group, no such collective idea space can be imagined between two individual idea spaces, then no meaningful social grouping can be imagined either that would place them together in a collective idea space. We can call the collective idea space taken at its highest level of aggregation the *foundational idea space*. The core principles, taken together, define the essential worldview of the social group, and we can call this basic worldview the *foundational worldview*.

It is illustrative to visualize the foundational idea space by making use of the hierarchy of fuzzy rules that we considered in the previous two chapters. In Fig. 4.1, we have created the hierarchy for some imaginary social group. The four core principles serve as labels for four fuzzy sets of ideas. Four distinct time periods are depicted as the horizontal dashed lines, and the depth of the hierarchy is shown as developing over time. We have made a number of assumptions in the representation, including on the almost identical triangular shape of the membership functions for ideas that may inhere to the core principles, the distances between the core principles, the evolution of their scope over time, their relative importance, and so forth. However, if we keep these

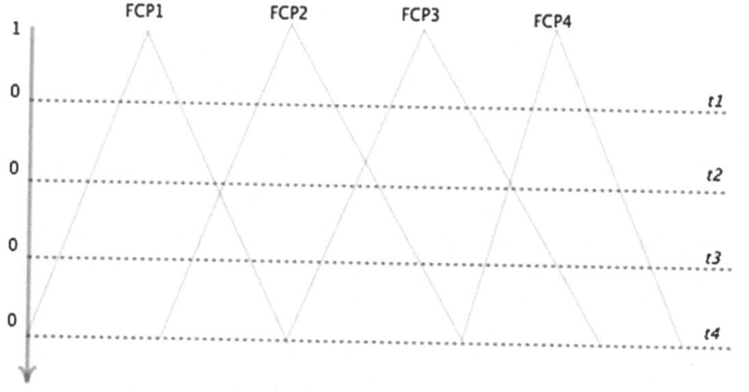

Fig. 4.1 A hierarchy of fuzzy rules from the foundational idea space

caveats in mind, the illustration does help us make some useful general observations.

It is worth noting that the scope of feasible ideas increases at an increasing rate; in other words, with the passage of time an ever-widening base of ideas can be associated with the core principles in the social group, which is a fact that sits comfortably with the infinitely expansible nature of ideas.

This proliferation of ideas permits an expanding set of feasible collective idea spaces to develop in the social group, yet, consider that, with time, the number of ideas that simultaneously inhere in each of the core principles also increases, which increases the relative appeal of such ideas.

Finally, note that a hierarchy of fuzzy rules for the foundational core principles evolving in this manner makes it more feasible for ideas and, therefore, collective idea spaces, to be developed less directly from the core principles. With the passage of time, most collective idea spaces, as a matter of fact, will only have an indirect basis in the foundational idea space.

4.3 Idea Imbalance

Another aspect of the foundational idea space is that the effects of the core principles are not uniform across space and time; specifically, these core ideas exhibit limitations to their crispness as the spatiotemporal connections between the overall idea spaces of individuals decrease. A consideration of the relevance of *propinquity* and *time inconsistency* is useful in capturing this nature of the core principles; both concepts serve to clarify the importance of essential similarities across individual idea spaces to the theory, whether they are based on similar ideas at any given moment across a group of dispersed idea spaces or a group of people who develop similar idea spaces over time as their idea spaces develop.

To help us consider both these claims, it is worth defining the concept of an *idea imbalance* more explicitly. In the theory of ideas, we may consider the degree of congruence that individuals have in their idea spaces with a particular collective idea space in a given context. At least three outcomes are possible here. If we are both members of the same group, say, a club, company, or community, your idea space may have a higher degree of congruence than mine with the collective idea space's core rules. Alternatively, your idea space may have a proportionately higher congruence with the collective idea space's crisper rules, or it may have a higher degree of congruence with the collective idea space's fuzzier rules. Since a group's core rules are inherent to its overall worldview, to the extent that your idea space intersects more completely with such rules than does mine, your advantage in this respect would translate into making you a more integrated and accepted member of the collective. This, then, has the effect of raising your propinquity with the group relative to mine.

Since an acceptance of a group's core rules is conducive to an individual idea space being compatible, any shortcoming in an individual's idea space in this respect would either prevent her from becoming a member of the collective idea space in the first place, put her at risk of being removed, or increase her chances of leaving; alternatively, the discrepancy may endure if certain core rules in the collective idea space are not integral to her idea space and, therefore, not routinely flouted. In the last eventuality—that of low and enduring propinquity—accepting the core rules does not entail an immediate disadvantage to the individual. Naturally, the larger the set of core rules, the more likely this source of idea imbalance becomes possible across the individual idea spaces. Being a citizen of a country may not seem disadvantageous to an individual who is willing to subscribe to all but a few of the country's rules and norms, especially when such rules play little or no role in her own idea space.

When the congruence of your idea space with the collective idea space is greater than mine on account of proportionately more crisp rules, you have an advantage in terms of being able to devise more new rules sequentially that leverage the mechanisms of the collective idea space; depending on the congruence of these new rules with mine, this situation of imbalance may serve to either further marginalize my idea space from the collective idea space (and, therefore, me from the group itself) or can serve to progressively include me within it more completely. In this latter case, my propinquity increases with the group through time, raising the level of congruence of my idea space with the collective idea space. As a result, my preference for the social group appears time inconsistent. Finally, if your idea space contains a higher proportion of fuzzier rules than mine that are congruent with those within the collective idea space, this would permit you an advantage in terms of the future trajectory of the collective idea space to the extent that those ideas become crisper in the collective idea space as well, or it may prove to become a relative disadvantage for you if they were to remain fuzzy.

Besides these broad characterizations of outcomes, one can imagine other sources for idea imbalance as well; however, the point here is simply to point out that this rather straightforward notion of an idea imbalance is at the heart of the four overarching core principles of safety, longevity, fairness, and control. A decrease in the imbalance between individuals in a collective idea space results in increases along all four of those aspects. However, these principles are at the apex of the hierarchy of fuzzy rules and are, consequently, rarely instantiated to the same degree directly in every instance of a collective idea space. Variations in context across collective idea spaces necessitate different degrees to which these principles can be achieved, how they are defined and operationalized in practice, which among them receives relatively more emphasis, and which is underemphasized.

To understand this more vividly, it is worth looking at contracts next. We shall argue that contracts can be seen as specific instantiations of underlying collective idea spaces at lower levels of aggregation than a foundational collective idea space, and that they hypostatize the core ideas as bases for their own mechanisms.

4.4 Variety in Contracts from Collective Idea Spaces

Before we can make clear what we mean by 'contracts' for the purpose of our theory, it is worth considering the range of contracts in practice. A formal contract can exist between any buyer and seller in the market. They may apply across an entire set of people, such as a union-negotiated employment contract, or they might be very specific in their duration, such as an options contract. Less formally, they can be broader in scale.

A constitution or treaty are examples of political contractual arrangements. And, several religious and cultural traditions interpose themselves as contractual understandings, such as with primogeniture or polygamy.

Perhaps the broadest category of contracts is that of legal contracts, which need to abide by the particular laws that govern their creation in order for them to be enforceable by a third party, usually the courts. One of their more ubiquitous and enduring requirements is that of 'consensus ad idem', which speaks to the degree to which any contract represents a consensus of the intentions of the parties to it at the time of its inception.

Such a shared consensus is obviously harder to justify for less formal contractual arrangements that operate at higher levels of aggregation—such as with social contracts in the contexts of citizenship (i.e. constitutions) or religion (i.e. doctrine)—especially when they also have durations that far exceed the lives of their creators. Since such contracts are not subject to frequent change, they cannot be informed by their immediate context.[3] Nevertheless, the argument on whether some shared morality or belief system emerges endogenously in the case of social contracts, or whether it requires explicit creation, is the broadest distinction on the political philosophy of the purpose of a social contract. What does seem very likely is that the inertial change we observe in the most enduring social contracts strongly suggests their basis in a set of foundational ideas that are shared across a community as direct expressions of the core principles.[4]

For a more contemporary example, consider the idea of smart contracts, where the shared intentions are entirely automated by way of a computer program, even immutably coded in their entirety on a blockchain. The ledger of transactions that a blockchain makes publicly verifiable to all its participants is essentially able to do away with third-party verification altogether. Since smart contracts eschew contextual variability as a source of information almost entirely, the immutability of a blockchain permits them to represent very large collective idea spaces, albeit only for entirely specifiable taks; instantiating ideas of participants directly

[3]Chiefly for this reason, Thomas Jefferson's contention was that the constitution ought to be rewritten every generation; his suggestion was duration of 19 years.

[4]Leeson (2009) provides an interesting discussion.

into smart transactions using the blockchain limits the scope of feasible transactions precisely because they can only apply to relatively crisp ideas.

Several informal contracts fall under another sweeping category comprising only verbal or implicit agreements. Usually, these characterize situations where the external environment is much harder to objectively specify, such as those situations involving honor, reputation, trust, or loyalty. The relevant collective idea spaces for such agreements are naturally much broader and, therefore, much more reliant on the core principles of the foundational idea space. Consider the practice of dueling, for example, which drastically reduced in its prevalence by the mid-nineteenth century, as it became increasingly possible to use the legal system to 'seek satisfaction' in cases of defamation.

A very different type of classification for contracts comes from contract theory in economics. Broadly speaking, in economics contracts are seen as being either incomplete or complete.[5] The study of incomplete contracts is inspired by the fact that, empirically, most contracts are verifiably incomplete. The motivation for why such contracts exist is usually an admixture of four observations: the parties are unable to come to an agreement about all aspects of their relationship; they are unable to describe the features of their relationship adequately for a third party to have any useful guidance in the case of a dispute; they are not able to foresee all the eventualities that may be immanent to their relationship; and finally, they may deliberately leave a degree of incompleteness in order to allow the contract to serve as an anchor for mutual expectations and to provide the parties some room for variance in performance. Note that all such justifications describe situations where a collective idea space relies very fundamentally on fuzzy ideas for new ideas to be consolidated and realized, usually through praxis creativity.[6]

A great deal of theoretical advances on the subject of contracts since the 1970s in economics have been made with the principle of

---

[5]The literature is immense. Some useful references that provide both summaries and advances are Hart and Tirole (1988), Maskin and Tirole (1999), and Hart and Moore (2008).

[6]Complete contracts, on the other hand, shift the focus to designing suitable mechanisms that can be implemented to elicit the true preferences of the parties involved and in describing the distributions of the overall states that might eventuate in the context of a relationship; that done, the aspects that justify the existence of incomplete contracts arguably vanish.

information asymmetry as a premise. The effort has been on clarifying the nature of differences in information across entities participating in a market, and what such differences imply in terms of the potential for markets to produce adverse outcomes for one group while benefitting another or, in the extreme, even cause the entire market to fail.

A variety of different mechanisms, all of which relate to information asymmetry in some manner, can conspire to create situations that are very pertinent to a contractual relationship. Moral hazard, for example, arises from the situation where one set of individuals in a market is afforded the incentive to undertake actions that carry an undesired indirect impact, or negative externality, on others. Since the cost of their actions does not accrue to them to quite the same degree as the benefit does, the incentives of such individuals have to be modified if they are to undertake changes in their actions that are favorable to other participants in the market. On the other hand, adverse selection refers to the circumstance where an individual withholds private information, presumably deleterious in nature, from others in the market. Since withholding such information provides the individuals with an unfair advantage, yet again they have to be given some incentive in order to motivate them to reveal it.[7]

The central idea worth reflecting on here is that there exists a common thread across all forms of contracts.[8] Contracts emerge in order to instantiate crisper versions of the broader foundational core principles—safety, fairness, longevity, and control—that are inherent in the core ideas of the collective idea spaces that they derive their context from; the more crisply and concretely they reify these core ideas of their participants, the better chance they have at enduring.

In this sense, the most general form of a contract is that of a collective idea space, and the most general form of information asymmetry is that of idea imbalance.

---

[7]It is the design of ingenious mechanisms that motivates this type of truthful revelation that complete contract theorists have helped specify.

[8]In Goorha (2018), an attempt is made to integrate a variety of contractual relationships that exist in a society within one holistic approach in order to define a context.

At its creation, a contract reflects commonly held information—the degree of congruence across idea spaces—regarding some initial set of ideas for its participants in a given context. These ideas suggest one or more mechanisms for the creation of new rules, the protection of existing rules, or both. A contract does this more specifically and formally than a collective idea space for a context that is of interest and that represents potential value to its participants under the compact of cooperation.

A multitude of challenges that contracts face arise from information asymmetry, which is, in effect, similar to the challenge a collective idea space suffers from the degree of the imbalances in the types of ideas between the idea spaces of its members. Such imbalances are obviously different from one collective idea space to the next and naturally depend on their context. However, they also ought to be seen as having a common source in the hierarchy of fuzzy rules that begins with the foundational core principles; context plays a variegated role precisely because every instantiation of a collective idea space relies on a collection of core rules as its basis, which are all drawn from this hierarchy, with variations that are sometimes only subtly different and at other times markedly so.

Similarly, what types of contracts are feasible or what features a successful contract eventually exhibits also depends crucially on a contract's resolution of the idea imbalance between its participants, under the shadow of opportunities that exist in the broader context external to its scope.

This is an observation worth some thought because it suggests why different types of contracts exist and, equally, why they may *not* exist. To the extent that cooperation yields some desired and specifiable value to a group of individuals, and to the extent that the external context represents opportunities that can prevent the value from being realized, or perhaps poses the risk of reducing it, an appropriate contract is desirable. The point to consider here is that the specifiability of this value in a contract often differs from context to context precisely because the collective idea spaces that define their mise-en-scène are themselves instantiated differently.

Consider a few different cases.

Imagine a collective idea space that represents a low level of idea imbalance—high congruence—across individuals over their core rules, a few crisp rules, and no fuzzy rules. In such a situation, a contract that seeks value not from the generation of new rules but from the preservation of existing rules has a greater likelihood of enduring by instantiating core ideas that are closer to the principles of safety, longevity, fairness, and control. Doctrinal ideas, be they religious or political, may be seen as traditional examples. However, a more modern example may involve systems of artificial intelligence with virtually no idea imbalance across the individual units that comprise it and with the value of the system emerging from its ability to preserve its instantiation of the core principles in its operations.

On the other hand, we may imagine the case where the idea imbalance in a situation arises from relatively high congruence on fuzzier rules relative to crisper rules. The situation is one where the core ideas may be very similar at the outset, but are more apt to being changed in unforeseeable ways as the collective efforts of individuals make some of those fuzzy rules crisper and introduce other fuzzy rules to the awareness of the individuals in the collective idea space. The abundance of fuzzier rules relative to crisp rules keeps such a collective idea space from being dominated by any one set of crisp ideas. Contractual value in such a case is expressly not based on the principle of the safety of crisp rules, but on the safety of the collective idea space it is based upon which favors fuzzier rules to begin with, regardless of whether they alter some existing crisp ideas. Combinatorial creative enterprises are premised on this form of collective idea space. Improvisation in jazz is an example of this type of collective idea space; jazz musicians pride themselves on expressing themselves creatively and being unbound by structure.[9] Yet, at the same time, the very idea of being iconoclastic is a core rule. Some genres of painting, like abstract expressionism, provide similarly interesting examples (Phillips 1974).

It has been argued that hacking is not dissimilar from painting in its emphasis on experimental creativity. Generally, open-source software development has been seen as an example of a collective idea space that emerges from co-opting individuals who seek an outlet for their creativity and cannot find it within the more structured firms and institutions of which they are members (Setia et al. 2012).[10] It has been argued that the organizational structure within firms—that is to say, whether the organizational structure is hierarchical, horizontal, or even 'spaghetti'—has a strong mediating force on the ability of a firm to exploit other avenues of innovation that are external to the firm (Foss et al. 2011).

---

[9]See May (2003) and Zack (2000) on this topic.

[10]Graham (2010) is an interesting investigation on this broad issue.

4.5 Cooperation as the Foundation of Competition

A rather important point now deserves our consideration. Our theory suggests that we ought to see competition not as the opposite of cooperation, but as an extension of its logic and based directly upon it. This is an observation that emerges naturally from our preceding discussion, and we can now state it more emphatically with the benefit of a few observations we have made along the way.

We begin by recalling that the core of any collective idea space represents those ideas that are crisp to all of its members and are crucial to the cohesiveness of the idea space itself; in this sense of cohesiveness representing stability, the core ideas, therefore, preserve a bias toward the status quo.

What can we say about the foundational collective idea space in this regard? We have suggested that the core principles undergird this collective idea space, applicable at the highest degree of aggregation. Since the foundational idea space is *the* collective idea space common to everyone in the broadest social group—in a sense *defining* the group—its core has a strong bias toward preserving the social group itself.

The four core principles combine to create the overarching worldview of all individuals in the social group that enables this task. The foundational worldview serves to enhance the foundational idea space's *safety* by preventing any other rules—be they fuzzy or crisp—from being accepted into the foundational idea space should they hold the potential to replace or revoke extant rules and violate the core principles. New ideas that reduce the *control* that the foundational worldview provides its adherents over the idea space are also resisted. Ideas that disregard or excessively alter the balance between the core principles pose the risk of undermining the notion of what is considered *fair* or permissible behavior according to the worldview of the core principles and are consequently avoided. Finally, any rules that might reduce or adversely impact the endurance or *longevity* of the foundational idea space are also spurned.

Together, then, not only do the core principles create an environment of stability founded on a bias for the status quo in the foundational idea space, but the premise that their worldview enables gives rise to the appearance of a broad umbrella of cooperation. Admittedly, the four core principles set a low bar for individuals to accept, and, indeed, it is easy to see that they can permit the admittance of starkly different, even incompatible, worldviews. We shall illustrate this idea more vividly in the following section.

Yet, their simplicity belies their efficacy in enabling cooperative outcomes at the level of the relevant social groups. The enduring successes of civilizations, societies, traditions of knowledge, religions, and institutions over long spans of times undeniably stand as compelling evidence for grand collaborative efforts. A range of mechanisms for these collaborative outcomes has been specified by disciplinary research, ranging from the invisible guiding hands of market price, divine destiny ordained by the numinous, the methodical pursuit of scientific truth, the visible guidance provided by an especially incisive line of leaders, or institutional restraints grounded in unimprovable political philosophy.

The overriding mechanism from our perspective, though, is simply that of *coordination in the defense against new ideas that oppose or threaten a commonly held worldview*. And what matters most keenly for our theory is the presence of cooperation in highly aggregated collective idea spaces that stand the test of time.

It is, of course, arguable that the nature of cooperation is rather weak and uncrystallized at the scale of a foundational idea space. The reason for this, logically anyway, is that we tend to think that people have little in common with most others in the largest social group that they belong to and can hardly be said to be actively 'cooperating' with them. However, this is a rather limited view (guided as it is by individuals rather than ideas) since it discounts the benefits that individuals collectively accrue from being able to implicitly rely on core rules that inform their worldviews and that transcend their immediate groups. In other words, core principles loom large in what one individual in a social group can take for granted about the behavior of another.

A surprising number of social norms have parallels in vastly different social groups, and their ubiquity has so immunized us to their efficacy that we tend not to grant the possibility that many of them are likely derived from common principles that enable us to function in our social group collaboratively. Yet, the manner in which each social group instantiates core principles creates differences in the worldviews across social groups, often irreconcilable, and these differences clearly suggest that foundational collective idea spaces are not immune to context.

At lower levels of aggregation, the collective idea spaces of social groups must contend more keenly with the 'context', which is itself a product of a variety of factors, including an amalgamation of particular geographies, the distribution of other collective idea spaces (what we can think of as the 'institutional environment'), and the distribution of other individual spaces (which is to say, the density of ideas among the social groups).

The focus on context that accompanies the collective idea spaces at lower levels of aggregation than that of the foundational idea space inspires a proliferation of mechanisms, which themselves arise as new ideas are realized in a variegated manner across the given contexts. Such collective idea spaces are much smaller than the foundational idea space, and, within them, there is now a keener presence of the appearance of competition, as new ideas are realized differently, sometimes with vastly different mechanisms and sometimes with identical ones. The awareness of ideas begins to differ across these collective idea spaces as the balance between the subspace mechanisms begins to differ markedly across contexts. And, therefore, new rules increasingly incorporate fuzzier ideas that exaggerate the contextual differences; this process of the innovation of new rules is a result of awareness that is distorted towards one set of mechanisms at the expense of another. The core principles that are at the basis of the foundational idea spaces never alter, the gradual pull of the constituent smaller collective idea spaces and individual idea spaces serving only to weight the four principles somewhat differently.

So, collective idea spaces in different contexts will differ from one another even if they all seek to realize the same initial given idea. Contextual differences will inspire them to vary in their use of mechanisms, and these differences will ensure that the particular combination of core, crisp, and fuzzy rules that the collective idea spaces will come to represent will also differ. To a degree, we should expect that the exchange of ideas between social groups should serve to remove differences in mechanisms and bring them closer together. And, indeed, this is an oft-cited effect of globalization and free economic and cultural exchange. However, we cannot expect all contextual differences to be forever removed, and given enough time and a reduction in the frequency of exchange in ideas, we would expect the variation to re-emerge.

4.5.1 Parable of Idea Spaces

Given our framework thus far, we can now construct a scenario that permits us to understand some key ideas of our theory. To do so vividly, let us permit ourselves some creative license and construct a story to provide us with the necessary context.

In the beginning, we can imagine an initial social group and call it the 'foundational social group'. Since our purpose here is to develop the theoretical import of our argument, we need not imagine that this social group actually corresponds to any specifically identifiable community or tribe of any kind from history. However, in order to fix the mind, it may perhaps help to imagine the foundational social group as a small set of hominins at the beginning of some recorded time.

The collective idea space of this initial foundational social group comprises just four core rules: safety, longevity, fairness, and control. These ideas form the fundamental worldview of the foundational social group, based directly on their most basic need to carry on and survive as a group. Let there be no other subspaces within their foundational collective idea space and no idea imbalance between the individual idea spaces of the individual members of the foundational social group. We might even imagine the initial set of individuals as clones of one another to develop this scenario.

We can now imagine that this group is split in two and that each subgroup heads off in a different direction before settling down again. Depending on the stories we wish to imagine, the rationale for splitting could simply be a desire to hedge bets against a dwindling resource, an environmental catastrophe at their place of birth, ostracism and exile, or perhaps just an amicable desire to explore some vast terrain collaboratively.

Regardless of their motives, we can perhaps imagine that each group ends up settling 10 miles from one another, one group on a mountaintop and the other in a valley. Or, perhaps, they end up settling 200 miles from one another, on the opposite sides of a sea. Our interest is really only in examining how their initial collective idea space is impacted by context differently.

Once the two new groups have resettled, we now confront their collective idea spaces with ideas that vary according to the exigencies of their new contexts. Initially, we may simply imagine these new ideas to emanate from some exogenous circumstances; for instance, there may be large differences in their geographies and climates, or in the types of resources they each encounter and employ. While each of the four core principles remains valid to the initial worldviews for both groups (they still want nothing more than to survive!), the type and rate of the information that the groups learn from their new environments now begin to differ.

Several effects are worth reflecting on. First, to the extent that one group systematically receives more information than the other—that is to say, the group benefits from a higher density of ideas—it is provided with the opportunity to develop more new rules than its counterpart. Perhaps one group experiences a far richer environment and range of natural resources and enjoys great variance in its flora and fauna, while the other group's environment is more homogenous, or perhaps it is little different from the place of their common origin.

Second, if the two groups receive information that pertains to one of the core principles systematically more often than the other core principles, their respective collective idea space will have more new rules that associate with that particular core rule, and, consequently, the group will develop a bias towards that core principle within its worldview. Thus, we might imagine that one of the groups begins to overemphasize fairness and longevity. This may be attributed to a range of circumstances. Perhaps the group enjoyed an environment that provided them with plentiful and varied resource abundance. These resources motivated the individuals to exert considerable effort in taming and employing them towards making things for themselves, but only if they would also be able to protect the future fruits of their labor from usurpation by other members. As a result, the members made commitments on fair treatment and long-term ownership. Meanwhile, the other group may have settled near a lake teaming with fish. Once they had settled in, they discovered that the lake was also a watering and feeding hole for hostile, man-eating animals. As a result, this group instituted rules pertaining to preserving their safety and controlling the threat of their shared danger in an orderly and organized manner.

While these might seem like overly contrived premises, they do allow us to highlight a few interesting issues. Since the four foundational core principles are, in a general sense, inviolable, whatever new information the group receives from their new surroundings does not yield rules that 'replace' any of the core principles. However, variety across contexts ensures that the core rules are clarified or refined differently by way of a series of contextualized hierarchies of fuzzy rules.

New information can quite conceivably serve to bias the group away from one of the core principles and towards another. This occurs simply because the dominant core principle sits at the root of several other new rules. As the relative number of ideas in the collective space that belong to the dominant core principle expands, more perspectives on any new ideas will emerge that employ the aspects of the dominant core principle. Even those new ideas that could be associated with the aspects of the weaker core principles get associated with the dominant one instead. When the resource availabilities of the first group alter, so that some remain plentiful while others become scarce, it becomes harder to separate individuals on the basis of how long or even whether they use the scarce resource, yet still preserve their worldview that has emphasized fairness. Similarly, when the man-eaters plaguing the second group have all be vanquished, it becomes hard for that group to dismantle their hierarchy of control. An unseen beast is, after all, far scarier and easy enough to invent.

The problem with such a bias is easy to see. So long as the context remains conducive, the bias goes unnoticed. After all, it assists the group's survival and even helps the collective idea space to flourish; several new rules are added to the collective idea space, some of these become crisper and suggest new mechanisms that form the basis for new subspaces. In other words, the social group begins to specialize on the basis of its entrenched worldviews. So deep can this entrenched worldview become that, when the context alters to one that favors one of the other foundational core principles instead, new ideas continue to be associated with the previously dominant core principle.

The import of this observation becomes clearer if we now imagine that our two groups again split in two and head off in different directions. The first branch from each group resettles in isolation, but the second ones come in contact with one another. The collective idea spaces of the interacting subgroups are now sufficiently different from one another, even though their ancestors had been clones.

Naturally, we cannot assume that the worldviews of the two subgroups will be similar because each subgroup may have developed subspaces within their collective idea spaces that are somewhat or even entirely new to the other, and crisp rules from these subspaces may have become part of their worldviews. An overemphasis on safety may have developed a subspace pertaining to ideas on war especially vividly in one subgroup, and combat may have become an integral part of their worldview. The other subgroup may have no such combative impulse. For them, an overemphasis on fairness and longevity may have permitted a subspace based on ideas of commerce to thrive, and mercantilism may have become an inherent part of their worldview. Given sufficient differences in context, and permitting enough time to elapse so that the differences that do emerge in the worldviews of the two subgroups become entrenched, it may even be conceivable that the groups find it infeasible to coexist, let alone cooperate.

However, the basis for hope that collaboration between such divergent groups remains feasible rests on their foundational core principles, which remain identical at their root. Stated differently, the apex of the hierarchies of fuzzy rules for their core rules is shared. While we cannot assume that the admixture of their collective idea spaces would restore the imbalance between the four core principles across both subgroups to some happy balance, we can certainly imagine that essential ideas that are associated with the core principles can be shared at a high-enough level of aggregation. Naturally, we might imagine especially strong resistance to some ideas, and the crisper they are for each subgroup, the more ardent is the resistance likely to be. Worldviews comprise core rules, which themselves contain mechanisms from subspaces. As such, resistance to certain forms of ideas embodied in essential theoretical and practical mechanisms (such as religion, science, heuristics, and technologies) is likely to be especially hard to reconcile when they are crisp for each subgroup (Mokyr 1992).

The eventual outcomes depend on a range of factors, such as whether one of them is conquered and whether the new context favors one or the other. Meanwhile, the other subgroups that resettled in isolation, or perhaps never left, are, of course, at the mercy of their context, and whether their collective idea space develops new rules that are conducive to their context once their peers have left them. Regardless, the world, as a whole, that all of our groups now live in is marked by a greater diversity of ideas.

4.5.2 Firms and Commons as Collective Idea Spaces

It is worth now revisiting the discussion we began in a previous chapter on what a theory of ideas can say about the differences that exist between *types* of collective idea spaces, especially when we permit each to represent distinct clusters of individual idea spaces as well as amalgamations of other collective idea spaces.

In order to do so, let us again consider the open access communities for sharing ideas—or knowledge commons—that we had briefly introduced before. For such knowledge commons, one might fix the mind on one of myriad enterprises that encourage relatively, or even entirely, free access to some repository of information. What is interesting is that these databases of knowledge are collated by individuals who voluntarily provide private knowledge gained through costly specialization. The motivations for such voluntary contribution stand in stark contrast with that other type of collective idea space, the firm.

In firms, by contrast, common sense and elementary economics suggest that the worker ought to be compensated in proportion with her contribution of labor, and that this should be expected to occur in close correspondence with the productivity of the worker and her contribution to the firm's value. It is worth mentioning here that there is a limited strand of literature in empirical economics which suggests how this benchmark is only dimly visible in real-world data, dulled by the extensive scope of underpaid and unpaid work done within firms, though this is usually attributable to less than ideal conditions in the labor market or regulatory environment that provide a bargaining advantage to the employer (Bell and Hart 1999). Regardless, both traditional firms and the open access knowledge commons can essentially be seen as examples of collective idea spaces with a set of core ideas that are derived, either directly from the broader foundational idea space or, more likely, indirectly through other collective spaces that operate at a higher aggregation than them. And, as collective idea spaces, they are both composed of the particular distributions of individual and collective idea spaces that they immediately rely on.

Yet, there are obvious essential differences between firms and knowledge commons, in addition to the variations in motivations of their members. Some of these can be usefully understood by examining the manner in which their *internal* worldviews correspond with the social group's *foundational* worldview. To see why, it is worth taking a brief digression and thinking about the diversity of both forms of collective idea spaces.

If we were asked to think of examples of knowledge commons around us, we would be tempted to include open online forums high on the list. There are, after all, a fair number of such forums on a range of topics, including physics, math, cuisines, astronomy, cars, home design, software, and sports. Indeed, given the preponderance of their instances that can be found at present, it would be quite understandable if one were to imagine that the very idea of knowledge commons is entirely a modern phenomenon, generated either directly or indirectly by the Internet and its attendant technologies that enable broad communication of ideas at low cost.

However, this is a bias that is chiefly a result of the 'physical separation' of individuals from their idea spaces that the Internet enables. This engenders a unique form of coordination and association between the idea spaces of individuals who then attempt to generate new rules and assess their realizability within the context of a collective idea space. It is perhaps tempting to add that the instrumentality of the Internet ought to be coupled with a second factor: a greater degree of tolerance for diversity in many societies than in the past, since it is a tolerance for diversity that seeds the vibrancy and enables the density of ideas represented in open access knowledge commons.

However, an acceptance of such diversity is arguably itself enabled by collective idea spaces at larger levels of aggregation that have core rules that are more compatible with the foundational core principles and, yet, espouse a worldview that does enable diversity. The fact remains that these factors create an easier environment for individual idea spaces to be associated based on idea balance in fuzzy rules and with lighter emphasis on the core and crisp ideas that describe the individual's worldview. This encourages more numerous, more diverse, and larger knowledge commons, and with the potential for more innovations as well.

There is, of course, a great deal of variation across firms, and there is a strong selection bias in comparing knowledge commons with all firms, since the latter comprise collective idea spaces that are not expressly formed with an intention for the generation of new rules, or indeed a primary focus that relates to creativity and innovation. However, when a firm does interest itself with creativity and innovation, its structure and organization very often display characteristics that mimic those of a knowledge commons more closely. For instance, research in organization theory has provided compelling evidence that diversity among the employee base in firms is associated with a greater potential for innovations (Yang and Konrad 2011). And, one of the most enduring facts on the abilities for innovation among firms is that of agglomeration, or collocation (Ellison et al. 2010).

The rationales for why collocation might provide a boost to innovative activity have been studied and clarified in economics ever since the 1920s. Generally, the explanations emphasize a cost-based approach, where a reduction in distance reduces the transaction costs associated with the exchange of physical resources and ideas. Positive externalities ensue as a dense network of firms develops that then serves as an attractor for other firms, investors, and talent. The essential point of difference between this view and the view espoused by the Austrian school of economics is with the treatment of knowledge as being an inherently dispersed good embodied in individuals, which is then brought together by market processes. With this premise, collocation of firms would naturally create an increased potential for innovation as distributed knowledge becomes more unified. This is a popular view in other disciplines as well, besides economics, such as with the history of science and organizational science, and is recognized as a significant factor in urban planning and foreign direct investment policy.

The theory of ideas we are considering permits us to consider the relevance of transaction costs and dispersed knowledge simultaneously quite intuitively on the basis of idea imbalances across idea spaces between *categories* of ideas: core, crisp, and fuzzy. The associations of idea spaces on the basis of core ideas require the prior specification of a worldview as the basis for selecting compatible idea spaces. This, however, leaves open any idea imbalances pertaining to fuzzy ideas, which do not closely inhere to a worldview; in such a collective idea space, the extent of the dispersion of knowledge would then be more closely related to fuzzy rules. It takes no great imagination, therefore, to understand the role of the elaborate hiring practices of firms that are guided by a desire to ensure a degree of idea balance across individual idea spaces on core competencies and values.

Since the realizability of ideas is an overriding objective even for the most innovative of firms, an emphasis on essential core and crisp rules is essential when the potential informational value that fuzzy rules represent is a consideration. It is worth recognizing that these core ideas for a firm, such as profit maximization or customer satisfaction, are immanent in the foundational core principles. Core ideas can be allied with relatively crisp rules pertaining to management practices, marketing strategy, production technology, and so forth, in articulating a unique worldview for the firm. If a multinational firm wished to produce a relatively homogenous product across different cultural and social contexts, it would then necessarily strive to remove all those sources of contextual influences that may serve to alter its worldview. For such a firm, the worldview becomes even more closely recognizable as one that is compatible with the foundational worldview based on core principles. However, fuzzy rules are very often generated from an awareness of other ideas, and controlling their generation and influence is far harder.

A knowledge commons, on the other hand, can afford to specify a worldview that also emphasizes fuzzy rules, since, apart from only a broad idea balance pertaining to the mechanisms that the knowledge commons make use of, a common worldview based on core rules across individuals is less relevant. Indeed, when the mechanisms are themselves not very crisp, the worldviews of knowledge commons are of necessity more inchoate than they are in a firm where mechanisms need to be much crisper.

There is another reason why, for a vast variety of knowledge commons, there is little need to select individuals on the basis of their wider worldviews. Access to a knowledge commons is more permeable than it is to a firm. Cooperation for knowledge commons, therefore, comes much more directly from the foundational worldview *and* the more particular worldviews that the knowledge commons reflect pertaining to the mechanisms immanent in their collective idea spaces. Historically, however, when the separation of individual idea spaces from the individual was not possible to quite the extent that it is in today's connected digital world, collective idea spaces would first need to ensure that common worldviews existed between potential and existing members who were contributing their individual idea spaces. Sorting on fuzzy rules was far less feasible. This goes a long way in explaining the strong backdrop of religious affiliation that accompanied the pursuit of pastimes as diverse as gambling, dancing, cooking, gardening, and a large number of sports as suggested by the histories of several civilizations all across the world.[11]

4.5.3 Power Laws and Idea Spaces

To conclude our discussion of the theory, let us consider power laws—a class of simple statistical relations that suggests how a relative change in one variable is described by a proportional relative change in a different variable. It has broad theoretical applications, such as Zipf's law, black swan events, and the Pareto principle (i.e., 'the 80/20 rule'), and has been shown to have a very wide variety of practical applications as well.

[11]See, for example, Dougall (2013) on sports.

Most of these applications highlight the fact that, in contrast to the benchmark normal probability distribution, a power law distribution permits extreme outcomes with far greater regularity, be they subjectively positive or negative. Among the better-known real-world applications of the power law are the relationship between city sizes and their ranks, the concentration of wealth in a society, the frequency of words used in various languages, the sizes of firms in an economy, the concentration of contributions made on Web sites, and the role that a few Web sites play in the structure of the overall Web.

In some situations, the concept of preferential attachment in a social network provides compelling insight on the mechanisms that might generate a power law (Barabási and Albert 1999). The idea is that new links in a network of individuals (or entities) are not formed entirely at random, but by a process that advantages those individuals that already possess more connections, or perhaps a process that mixes random link formation with preferential attachment (Simon 1955). There are some plausible reasons for preferential attachment in a network. Economies of scale can certainly play a strong role in advantaging larger members of a network. The location of some members within a network can provide those that are connected with them with crucial direct and indirect benefits, such as reputation, access to resources, safety, and so forth. Such network effects can also play a role in generating preferential attachment. Homophily, or the preference to connect with other entities or individuals of the same type, will similarly yield a network that shows preferential attachment (Jackson and Rogers 2007; Mark 2003).

Our theory suggests a reason for why we might observe the stubborn ubiquity of the power law across a multitude of different contexts based quite simply on the structure of idea spaces.

There is a range of factors that we might consider when examining why preferential attachment might occur in the setting of some network where the entities can be thought to represent idea spaces. At the level of a foundational idea space, all entities in a social group can be said to be connected; in an environment of idea spaces that each reflects a worldview derived directly from the core principles, there is little reason for there to be any bias in the structure of connections between entities. However, with variations in context, the hierarchy of fuzzy rules becomes deeper as well as more variegated. The ability for idea spaces to associate with one another now begins to depend on ideas that represent particular perspectives or 'interpretations' of core principles, as well as the mechanisms available to cohere such perspectives into aspects.

To see this, we can imagine any two idea spaces, A and B. The ability for A and B to be associated depends on the degree to which there is an idea balance across their core and crisp ideas for a given context. Naturally, it is feasible that this idea balance between them could alter if we were to consider a different context. Now, imagine if one of these idea spaces, say A, also represents new ideas, but that idea space B does not. These new ideas need only have the feature that they be based on one or more mechanisms derived from the core and crisp rules that both idea spaces shared. To the extent that they do, the new ideas in A would yield perspectives that have at least some aspects that are shared between both A and B; it may even be that all such perspectives on the new ideas in A also belong to aspects that are common to B. Why would the new idea then not simply also appear in B as well? In other words, what prevents the individual with the idea space B from simply 'stealing' the idea from A merely by becoming aware of it? There could be a few scenarios.

If B has awareness of all the necessary ideas and mechanisms used in the generation of the new idea that idea space A represents, then, trivially, B could simply mimic A. If B does not represent all the mechanisms, but does possess all the ideas, then clearly the idea cannot be replicated unless the subspace mechanism, or metarule, is learnt first. This is similar to a situation where an individual has a sense for all the ingredients required for a dish she likes, but not the precise recipe for combining them.

Possibly, A may be relying on mechanisms from subspaces that B simply does not have access to; this would result in missing perspectives and aspects on the ideas in B. Alternatively, B may have access to all the requisite mechanisms, but may not have an awareness for all the ideas. This would correspond to a situation where a key ingredient or vital step in the recipe is missing.

In either case, the necessary ideas or mechanisms may be accessible, provided a sufficient reserve of awareness exists to access them.

Given the number of missing components in B, the necessary reserve of awareness will naturally vary. Finally, B may not possess all the ideas or all the mechanisms, or even the necessary reserve of awareness, and in such cases mimicking the new idea is quite trivially infeasible.

References

Acemoglu, D., & Robinson, J. (2013). *Why Nations Fail: The Origins of Power, Prosperity, and Poverty*. New York: Crown Business.

Barabási, A.-L., & Albert, R. (1999). Emergence of Scaling in Random Networks. *Science, 286*, 509–512.

Baumer, E., & Martin, K. (2013). Social Organization, Collective Sentiment, and Legal Sanctions in Murder Cases. *American Journal of Sociology, 119*(1), 131–182.

Bell, D. N. F., & Hart, R. (1999). Unpaid Work. *Economica, 66*(262), 271–290.

Dougall, A. (2013). Bad Sports? Puritan Attempts to Ban Games in 17th-Century England. *BBC History Magazine*.

Ellison, G., Glaeser, E. L., & Kerr, W. R. (2010). What Causes Industry Agglomeration? *The American Economic Review, 100*(3), 1195–1213.

Elster, J. (1988). Economic Order and Social Norms. *Journal of Institutional and Theoretical Economics, 144*(2), 357–366.

Foss, N., Laursen, K., & Pedersen, T. (2011). Linking Customer Interaction and Innovation: The Mediating Role of New Organizational Practices. *Organization Science, 22*(4), 980–999.

Goorha, P. (2018). Contextual Contracts: On a Context-Sensitive Approach to Contract Theory. *Journal of Interdisciplinary Economics, 30*(2), 1–19.

Graham, P. (2010). *Hackers & Painters: Big Ideas from the Computer Age*. Sebastopol: O'Reilly Media.

Hart, O., & Moore, J. (2008). Contracts as Reference Points. *Quarterly Journal of Economics, 123*(1), 1–48.

Hart, O., & Tirole, J. (1988). Contract Renegotiation and Coasian Dynamics. *Review of Economic Studies, 55*(4), 509–540.

Jackson, M., & Rogers, B. (2007). Meeting Strangers and Friends of Friends: How Random Are Social Networks? *The American Economic Review, 97*(3), 890–915.

Leeson, P. T. (2009). The Calculus of Piratical Consent: The Myth of the Myth of Social Contract. *Public Choice, 139*(3/4), 443–459.

Lipset, S. M. (1959). Some Social Requisites of Democracy: Economic Development and Political Legitimacy. *American Political Science Review, 53*(1), 69–105.

Mark, N. P. (2003). Culture and Competition: Homophily and Distancing Explanations for Cultural Niches. *American Sociological Review, 68,* 319–345.

Maskin, E., & Tirole, J. (1999). Unforeseen Contingencies and Incomplete Contracts. *Review of Economic Studies, 66*(1), 83–114.

May, L. (2003). Factors and Abilities Influencing Achievement in Instrumental Jazz Improvisation. *Journal of Research in Music Education, 51*(3), 245–258.

Mokyr, J. (1992). Technological Inertia in Economic History. *Journal of Economic History, 52*(2), 325–338.

North, D. (1994). Economic Performance Through Time. *The American Economic Review, 84*(3), 359–368.

Nowak, M. (2006). Five Rules for the Evolution of Cooperation. *Science, 314*(5805), 1560–1563.

Phillips, D. (1974). Understanding Jackson Pollock and Blue Poles. *The Australian Quarterly, 46*(4), 91–96.

Setia, P., Rajagopalan, B., Sambamurthy, V., & Calantone, R. (2012). How Peripheral Developers Contribute to Open-Source Software Development. *Information Systems Research, 23*(1), 144–163.

Simon, H. A. (1955). On a Class of Skew Distribution Functions. *Biometrika, 42*(3–4), 425–440.

Yang, Y., & Konrad, A. M. (2011). Diversity and Organizational Innovation: The Role of Employee Involvement. *Journal of Organizational Behavior, 32*(8), 1062–1083.

Zack, M. (2000). Jazz Improvisation and Organizing: Once More from the Top. *Organization Science, 11*(2), 227–234.

Part III

Why We Play

5
The Ideas Before Sports

It is perhaps uncontroversial to suggest that the context of a society matters a great deal to the fate of its sports.

In England, a royal decree in 1365 forbad playing a list of games that included handball, loggats (which involved throwing sticks at a stake), and football, simply because they diverted the attention and frittered time away from training at more useful pursuits, such as archery. The Book of Sports, first issued by King James in 1617 to counter the Puritan distaste for Sunday revelry, laid down the law of the land during the period prior to the English Civil War; it outlawed the play of only certain sports, such as wrestling, bowling, and bullbaiting, permitted others, such as archery and 'leaping', but then only upon the conclusion of church services on the Sabbath. This time the pretext was almost certainly King Charles's strong bias against certain religious sects. The law was ardently petitioned against by several of his subjects who yearned for more latitude in being able to pursue their sporting interests without giving unintended offense. The fact is that one of the first sports that has been recorded as having been played by the new settlers in Massachusetts is that of stoolball, which was perhaps a fortuitous

P. Goorha and J. Potts, *Creativity and Innovation*,
https://doi.org/10.1007/978-3-319-94884-3_5

respite for the new arrivals, since the English Civil War gave lease to the Puritans to ban *all* sports for a period shortly thereafter.

Besides the looming role of the wider context to ideas, sports are an especially pertinent subject for this book for a number of other reasons.

As group pursuits, sports rely on the *collective* idea spaces of their participants and followers; indeed, the visible features of a sport can be seen as a defined set of realizations from collective idea spaces for social groups. To the extent that this is true, we ought to be able to see evidence in these features, across a wide range of sports, of a range of antecedent ideas that are instantiated as core rules in the collective idea spaces.

Like any idea space, the collective idea space that a sport represents is influenced by the broader social context that it is played within. Often we will note uncanny similarities across different sports in the same context or similarities in sports across vastly different contexts that are separated spatially as well as temporally. In such cases, we ought to be able to appeal to a hierarchy of fuzzy rules for sports, rather than having to rely on some process for the diffusion of ideas through direct contact alone; using such a hierarchy, we ought then to be able evaluate the instantiations of broader core rules from collective idea spaces at higher degrees of aggregation in the core rules for the particular collective idea space of a sport and assess how they engender those observed similarities.

In this chapter, we set out to explore these issues, admittedly by taking a series of useful digressions that come together to suggest a larger picture that conforms rather well with the theory of ideas we have developed. It bears noting that we are not outlining a history of any one sport, concerning ourselves with its greatest players and chief patrons, or even unpacking the intricacies of how they are played. Our interest is broader; we will take the lunar view and focus on using sports as an exemplar for the theory of ideas and see how far we get by looking at their richness from that wide lens.

Before we begin, it is worthwhile reflecting on the fact that any given collective idea space provides an inwards *and* outwards reflection of its ideas to varying degrees. Inwardly, it comprises at least some commonly held components of the idea spaces of all individuals that are members of it; outwardly, it reflects in its structure those components that it shares with the collective idea spaces of groups at lower as well as higher levels of aggregations.

So, given the ubiquity and longevity of sports, we ought to be able to find traces of the obverse of our opening statement—that of the influence that sports exert on their contexts; indeed, it does not appear to demand any unreasonable stretch of the imagination to see that the individual idea spaces of members of a sport routinely impact the nature of their society, just as they are influenced by the societies they live in.

Sports clubs and teams are important components of civil society, though some evidence suggests that the role they play in connecting other organizations in society is debatable (Seippel 2008). Nevertheless, their efficacy in developing bonds among players is broadly acknowledged. The desire an individual has to cooperate with another member of her social group who also plays her sport, let alone belongs to or supports her team, is naturally a great deal higher than if the opposite were the case.

The greatest rivalries in sports matter deeply to the fans of the teams that are involved: New South Wales vs Queensland in Australian rugby; the Boston Celtics vs the Los Angeles Lakers in basketball; India vs Pakistan or Australia vs England in cricket; Argentina vs Brazil in international soccer; USA vs Russia in ice hockey; Barcelona vs Real Madrid in club soccer; the Boston Red Sox vs the New York Yankees or the Los Angeles Dodgers vs the San Francisco Giants in baseball; Ohio State vs Michigan or Harvard vs Yale in college football; the New York Giants vs the Philadelphia Eagles in American football. What makes each of these, among scores of others, so enthralling is in no small measure a result of the fervidness of the fans that barrack for their teams. You might forgive me more readily for supporting a rival brand of a product to the one you favor, but you may not find me conducive company if I didn't think much of the sport you favor, or perhaps worse, support the nemesis to your favorite team. Indeed, some interesting research in management theory, based on examining enduring rivalries in sport, suggests that the very idea of competition is itself grounded in 'relational attributes'; the personal, social, and psychological investments made by individuals in a sport or a team have enduring effects on the nature of its rivalries and the quality of performance (Kilduff et al. 2010).

The 'membership profiles' of individuals within a larger group—which is to say, the organizations that an individual belongs to—have often been seen as being predicated on the membership profiles of others that one admires, follows, or emulates. However, there is also a stickiness in certain memberships that comes from the associations of one's parents and grandparents. And, to the extent that others in a society also adopt similar guidance for their behavior, the probability that a group of people with similar membership profiles develop at least some loose bonds with one another grows. We shall see that, with sports, the strength of this sort of group affinity has a deeper meaning that has roots in the core ideas within the broader collective idea spaces of the social groups, which the sports rely on themselves.

It is worth remembering that, from the perspective of our theory, ideas recognize no discrete boundaries. There is no flag that tells us that we are leaving the bounds of the idea space of any one individual and entering the collective idea spaces of larger and larger groups of individuals, or vice versa.

This observation has a number of implications for the case of sports. Crucially, while the barriers to professional sports may well be very high, the barriers to participating in one or more sports *at some* level are a great deal lower than they arguably are for other groups in society pertaining to, say, political institutions, financial markets, and organized religions. As a matter of fact, we shall see that there is reason to think that the core of the collective idea spaces for sports may well be drawing upon innate human behaviors, which is as we should expect if the core ideas in a sport are based upon those drawn from larger collective idea spaces. This naturally enables a more fluid connection across the ideas spaces of individuals that may belong to different collective idea spaces at varying degrees of aggregation.

It is interesting to note that sports economists usually model sports teams as seeking to maximize wins rather than profit directly, suggesting that they understand that an organization's success in a sport cannot be built on the cold hard considerations that may drive a calculating (albeit, usually theoretical) firm in a market (Tremblay 2009). There is a distinct feeling that there are other dynamics at play in sports. While it is certainly true, for example, that the competitive trade off that teams within a league face on the acquisition of talented players is rather direct, in the larger context of a society and the longer-term prospects of a sport league's viability within it, this trade off ceases being quite as overtly competitive (Dakhlia and Pecorino 2006). Thus, the sport serves at least two social purposes. In the proximate context, it encourages competition across teams in the pursuit of talent and wins, yet, in the larger context and over a longer term, it encourages an environment that rewards cooperation in the promotion, if not the direct development, of the sport. This tension between the processes of competition that a sport displays at an individual or smaller collective idea space and the apparent cooperation that it favors at the scale of larger collective idea spaces makes sports an especially interesting case to examine (Kaplan and Hill 1985).

5.1 The Foundations of Sports

Rarely is a game 'just a game'. Even when played in the backyard between siblings, a friendly game always holds the potential for turning extremely competitive and quite charged with bubbling rivalries. And when played on the world stage, a game can change the course of history.

When, for instance, the American boxer Joe Louis overpowered his German rival Max Schmeling in June 1938, it was far more than just a bout. Louis was avenging the prior loss he had suffered to the German boxer; he was answering the call of his nation—and a personal entreaty from the US President—to quash the Nazi German claim of Aryan superiority that had accompanied their prior meeting. And, if that was not quite enough, he was also carrying the hopes of scores of his fellow black Americans who wished to see him triumph over a white man, to some degree regardless of what nation he sprung from.

So, perhaps, it is worth considering the broader 'purpose' of sports to a given social group. At the broadest level of aggregation, all sports can be seen as comprising some rules that are derived from fuzzy sets of ideas with labels that collectively serve to define this overarching purpose. Purpose is imparted to the sport by its participants and its spectators; it is they who determine what lies within the scope of the sport's intention and what transgresses it or falls short of fulfilling it. Purpose is at the crux of any sport; it is at the crux of the worldview of its participants, and, consequently, it is also related to the core rules at the heart of its collective idea space.

From the perspective of our approach, the ideas of various sports, as they are embodied in their respective collective idea spaces, can be seen to emerge from successive attempts at creating subsets of rules within a broader fuzzy set applicable to sports. Such subsets emerge as individuals, or groups of individuals, attempt to provide narrower ranges for one or more constituent rules for some idea within the broader fuzzy set. As we saw in Chapter 2, such attempts create a hierarchy of fuzzy rules. Often, an entirely new characteristic to the sport might be introduced that was absent in previous sports, which is then associated with some extant rules within the broader collective idea space for a wide range of sports. In time, the new rule may also be refined and included in the collective idea space of one or more sports as a crisper rule through the creation of its own hierarchy of fuzzy rules; alternatively, it may remain fuzzy and be discarded.

Sports have features that can give us fascinating insights on our theory's basis of seeing competition and cooperation as compatible concepts along a continuum, rather than as opposites. In this regard, we are interested in examining what is unique about that set of ideas that ties different sports together.

Perhaps one avenue to pursue in the search for these foundational ideas is to be found by looking at the studies on ludic diffusion—that is to say, the spread of sports across cultures and countries. Such investigations have examined the channels by which sports are introduced and become popular in a society. Cultural imperialism is often high on the list.[1] When one sees the intense media attention that some international sporting events generate, it is obvious that sports can indeed become hostage to the cultural, political, and social machinations between the countries that organize and govern a sport on a global scale. National pride is very routinely at stake.

[1]Guttmann (1996), for example, is a review favoring the cultural diffusion viewpoint for cricket.

However, the inference that culture is the chief determinant of ludic diffusion may not be very reliable, not least because it then shifts the problem to finding ideas inherent in different cultures that can usefully serve as the foundational set of ideas for sports. It is baffling, for instance, to understand why the logic of cultural imperialism holds for one sport but not the other, especially in those cases where both have demonstrably originated from the same country.

Idiosyncratic societal characteristics are often just as important as culture, though the two are prone to being conflated. For example, some prevailing explanations for why cricket did not thrive in the USA after the Civil War suggest the reason to be less attributable to its obvious associations with England, and more to do with the fact that it was seen as largely a pastime of the wealthy in Philadelphia and New York (ironically, just as this was becoming *less* of a truism) (Pauketat 2009). The first international sporting contest, ostensibly of any sort, was a cricket match played in 1844 between Canada and the USA, which makes it especially interesting that cricket did not take a hold in either country to anywhere near the same level of enthusiasm accorded to other sports such as baseball and ice hockey. A possible reason for this is the associations that the game carried to the English upper class. As such, it appears to have been played in clubs and private schools in both countries by members of the upper echelons of society who wished to distinguish themselves as 'gentlemen'.

To be fair, some studies have made the point that, while sports may originate from one country or another, to say that their diffusion and uptake is a sign of intentional cultural hegemony is perhaps too stern and unjustifiable of a conclusion (Riess 1994). They can also serve as a proxy for airing out political and social grievances. After all, unless entirely rigged (and often even then!), a match often provides the beguiling opportunity for a weaker opponent to vanquish a stronger one, leveling the playing field, as it were; the prospect of using the sports field to humble a greater power has no insignificant appeal.

If political, cultural, and social determinants are not reliable predictors for the cross-contextual appeal of a sport, the inference then is that there is something else that is intrinsic about the purpose of sports that enables them to be adopted by varied social groups. And, that too, in a manner that strongly suggests that a 'diffusion-like' process might be at play.

Our contention, derived directly from the theory we have outlined in previous chapters, will be that sports generally arise from the foundational idea space of a social group and, therefore, mimic those core principles within their own collective idea space. The generally unacknowledged fact is that the core principles of foundational idea spaces are more generally shared across individuals than a culture is shared among its own members. This permits the probability for the 'diffusion' of a sport across varied social groups to be internally driven and without the social group feeling that the sport has been foisted upon. When it is accepted, a sport is adopted by a social group on the basis of commonly held core ideas, and when it is rejected, it is on the basis that the core ideas of the sport stray too far from those core principles. To this, we add the observation that some sports permit opportunities for the introduction of a deeper rule hierarchy for its constituent rules, thus enabling them to become more realizable within a social group in different contexts than do other sports that have shallower hierarchies of rules.

5.2 Core Rules for a Class of Sports

Given our four core principles of the foundational idea space—safety, fairness, longevity, and control—we can now attempt to consider the core rules in a collective idea space for a class of sports that permits them to be acceptable to a social group purely on the basis of their inherent idea balance with those core principles. We call it a 'class of sports' to explicitly acknowledge that one or more broader collective idea spaces may exist from which, through a hierarchy of fuzzy rules, this class of sports may be developed. Nevertheless, we attempt to capture as large a group of sports as we can within our class in order to examine the degree to which its own predominate rules can be seen as compatible with the collective idea space for a social group at a broad enough level to make it of foundational importance.

On the one hand, this may seem like an impossibly complex task, requiring us to be well-informed about the nuances of sports that have survived and those that have perished over time across the globe; the task is complicated further by the fact that our recorded history of sports is rather limited even across the ancient civilizations, and virtually non-existent for vast periods of prehistory before then, thus biasing such investigations in favor of more recent sports.

However, our ambition here is less ambitious than to provide a historically accurate account of a variety of sports and a compendium of their rules of play. Instead, it is to see how a class of sports, as a broad category of social activities, can be examined from the perspective of our theory. This makes our task a great deal easier, since we can simply work from 'first principles' and contrast the core rules of sports with the core principles, albeit by finding their relevance in ideas that are derived from a social group's broadest context rather than the sport's specific rules of play, which are representative of a comparatively smaller collective space. This is an important point. Rules for a broad class of sports must be derived from a collective idea space at a higher degree of aggregation that is both appreciably smaller than that of a foundational idea space and, yet, linked to the intentions of its core principles.

It perhaps requires no stretch of the imagination to suggest that a sport cannot survive long if its constituent rules, defining its purpose, deliberately make it unfair, unsafe, or unwieldy. However, what we are at liberty to posit on the basis of our theory is only that, no matter how unfair, unsafe, or complex a sport may seem, it can endure within its social context so far as its core ideas do not have an idea imbalance with the core rules of the larger collective idea space that comprises the individual idea spaces of its various participants, and who are relevant to the consolidation of its ideas.

The more broadly a sport is played, the less the imbalance there ought to be between the core ideas of its collective idea space and the core ideas of the larger collective idea space of the social group to which it belongs. If this were not the case, the core rules of the sport would appertain to a worldview that would be in conflict with the worldview of the larger collective idea space in the same context, with the result that it would be less likely to endure. Either its rules would be amended to make its worldview compatible, or the core rules of the larger collective idea space of its context would have altered to suit it.

Since ideas are consolidated across a greater number of individual idea spaces in a larger collective idea space, the core rules and worldviews for larger collective idea spaces are more durable. Consequently, it is far more likely that the rules of the sport's collective idea space would need to be amended. Conversely, when a sport is played more narrowly within the context of a social group, its collective idea space comprises core rules that can be consolidated and realized by fewer individuals, representing a much narrower set of individual idea spaces. Thus, an imbalance between the core ideas across the collective idea spaces might be more sustainable in such cases.

It is worth recalling the manner in which a constituent rule that belongs to the collective idea space of a sport or set of sports might be adjusted to suit a particular context, giving rise to new rules and the concomitant possibilities for more sports to emerge. The two modes for rules changes are either through a revision of an extant rule or through the reinterpretation of it in a manner that reduces the idea imbalance. Specifically, consecutive clarifications of the rule might be proposed, and a hierarchy of fuzzy rules thereby developed, for the rule by one or more members of the social group until the rule can be adjusted or relabeled in a manner that reduces its imbalance with the larger collective idea space. Alternatively, an entirely new characteristic can be introduced to the sport that permits extending the intention of the extant rule or generating a new category of rules altogether, both done with a view to develop a new sport that enables a greater balance of its core rules with the worldviews of the larger collective idea space.

5.2.1 Four Core Rules for the Class

With this broad ambition of capturing a relatively large group of sports for analysis, we might begin by defining our class of sports as those that involve teams, an arena of play, an object to strike, and an object to strike with.

Our interest is in understanding what the core rules of the broadest collective idea space for this class of sports might be to enable it to serve as the basis for the collective idea spaces for a variety of other sports. All sports within this class should feature at least some core rules that have been developed through a hierarchy of fuzzy rules that begin with one or more of the core rules of this general class. In addition, we will try and trace the relevance of the core rules of this class of sports back to the core principles of the foundational idea space and to some collective idea spaces that arguably operate at a higher degree of aggregation than it.

Our definition for this class of sports suggests four core rules:

- *aiming and throwing (or launching) an object;*
- *hitting or striking with an object;*
- *arranging individuals into teams and organizing their placement on a defined territory; and*
- *a method for adjudicating an outcome.*

We explore the argument in this chapter that these rules can be considered core rules for a collective idea space for a broad class of sports since they have a firm basis in the foundational core principles themselves. It is evident even at the outset that these are rather broad rules and would permit significant variation in the types of collective idea spaces they can encompass, based on the relative emphasis placed on each of the rules and the depth of the hierarchies of fuzzy rules they can engender in any given context.

For example, the ability of an individual to hit a target effectively and with repeatable accuracy is, to a significant degree, determined by the differences across the objects that are thrown, as well as by the mechanics of the body involved in the process of throwing the object. One need only contrast the game of darts with that of hammer throw. Both involve launching projectiles that are very different, using mechanics that are nothing alike, yet they both are based on the general core rule of throwing, which is itself, we shall see, derived from a combination of the core principles of safety and control.

On the other hand, when a rule on being able to repeatedly strike an object is emphasized, the core principle of longevity becomes germane as well, in addition to those of safety and control. The implement used for striking can of course vary widely as well. It may simply be the open hand, as it was in the earliest forms of tennis; a closed fist, as it is in boxing; or a bat, like it is with baseball. Core rules on the makeup of teams and their placement on a territory borrow aspects from the core principles of fairness and control. The composition of the team can, of course, vary from a single individual, as it is with, say, snooker played on a 10-ft-long table or several hundred, as it was with la soule, a French game that seems to have been much like soccer for entire towns competing against one another, and with very liberal rules on what the object was for striking and what might be used to strike it with. Finally, the objective of a core rule on a methodology to adjudicate success in sports—chiefly those that pertain to scoring and officiating—is to assist in the determination of the capability and talent of individuals and teams. Core rules serve to provide objective standards that can be shared across individuals participating in the collective idea space. Here, quite obviously the influence of the core principles of fairness and longevity can be seen.

In addition to the four core rules we have selected and accorded the distinction of being necessary to a foundational collective idea space for a large class of sports, one may argue that we might have considered other ideas as well. Perhaps ideas pertaining to cognition, fitness, nutrition, and enjoyment seem to be just as crucial to the class of sports we are examining, and, as such, should also be seen as core rules relevant to a foundational collective idea space for sports. While this may well be true, our objective here is to examine the logic of our approach for this application to a class of sports by only considering a minimum set of core rules, and these other rules arguably cast the net too broadly to be of indispensable relevance to our analysis here. The sports that this overarching collective idea space of sports covers, regardless of their appearance or state of play, would have certain core rules as their most recurrent and foundational feature that draw upon one or more of the predominate rules from the broader collective idea space.

We will try and justify the roots of these assertions in this chapter more fully, but a relatively straightforward source for the intuition comes from studies on cooperative hunting, which have shown that the size, geographic concentration, and relative abundance of a prey species are all vital determinants of whether cooperation emerges among a group of hunters (Packer and Ruttan 1988). In safe-haven games, such as cricket and baseball, a member of one team is on the field facing all members of the opposing team. To the uninitiated spectator, this imbalance may seem patently unfair. However, seen in the context of a hunt, the team only acts cooperatively to vanquish the batter precisely *because* it is all against one. The batter, on the other hand, faces off against everyone in a struggle for individual survival, which goes a long way to justify the relatively greater emphasis batters place on personal accomplishments and targets of achievement.

There is, of course, a risk in taking this analogy too far without restating our assumption for why it may be valid to make the comparison: the foundational idea space provides core principles that enable cooperative behavior across all the collective idea spaces at lower levels of aggregation than it within a social group; the purpose of core rules in a sport that requires cooperation is to simulate those conditions.

5.3 The Very First Collective Idea Space

For some, a strong correlation between war and sports in a society may hardly be a difficult idea to credit.[2] Combat in war seems to feature all those core rules that we have suggested the foundational collective idea space for sports features: striking opponents, aiming and launching projectiles, organizing in teams to defend a territory and a method for declaring victory.

Indeed, history is replete with examples of warriors learning the skills of archery, wrestling, horseback riding, swordsmanship or tilting at quintain from an early age. Sports based on the skills needed in warfare can readily be seen as activities that are necessary pastimes in preparation for war. There is no dearth of examples of monarchs and rulers across the world declaring the merits of teaching the younger generations the art of warfare through simulated combat, blurring the line between sport and training for war. Javelin throwing in Greek antiquity,

[2]See Carter (1985) for an interesting review of this connection in the context of medieval feudal societies.

for example, was a sport practiced, with subtle differences, by both warriors and athletes (Harris 1963). The Cahokian game of chunkey, from a little more than a thousand years ago, featured vast teams of Native Americans throwing spears and mud balls at a rolling disk while running at full pelt (Speck 1944). There is a basis in cognitive psychology for the transference of mental skills across physical abilities,[3] so it is not merely in training the body physically that may have inspired the connection between war and sport, but very likely in training the mind of a potential warrior as well.

While, prima facie, this connection between war and sport seems promising, it really only suggests the possibility that the foundational collective idea space for *both* activities might be shared.[4] It does not, however, suggest very convincingly that the collective idea space for war ought to be seen as foundational to sports. One obvious reason for this skepticism is that it is far less evident for a range of sports that they have anything to do with war directly. And, more to the point, if we are interested in searching for the precursors of the core ideas that are shared across individuals in a group that are then realized in the features of a variety of sports and in war in different parts of the world, we ought to look further back at the dawn of human history on Earth. It would behoove us to reflect on the most basic of social activities, since they would be more directly allied to the foundational collective idea space of any emerging social group. We ought, in short, to extend our gaze far beyond the handful of millennia we feel most comfortable with when thinking of human history and consider the first principles of existence that then lent themselves to common core beliefs.

Perhaps, then, we might consider whether the core ideas in sports derived from the advent of primitive hunting more directly than from organized war, since hunting arguably has just as long a history in the evolution of human society as almost any other activity imaginable.

---

[3]See Jessup (2009) for example.

[4]Especially, notable is that fact that the origin of warfare in hominins is far from a settled issue. See Thorpe (2003) for a discussion.

At well over three million years, the Paleolithic era, or the Stone Age, spans a space of time that is preponderant over the course of our evolutionary story. During this time, several fascinating events occurred that are well beyond the narrow scope of our interest here in ideas. However, one aspect is particularly instructive: the technological transformations that took place in the making of stone tools during the Paleolithic. The overarching relevance of stone toolmaking, or lithic technology, is not only that it can be seen as an example of one of the very first, if not *the* first, collective idea space, but also that, as the nature of this collective idea space changed over time, it became more conducive to a collaborative activity like hunting.

That the imperative to make stone tools arose from a basis in every one of the core principles of the foundational idea space is rather plain to see.

First, they enhanced the safety and assisted in the longevity of the group in their ability to enable easier access to food, clothing, shelter, and necessary resources. Second, since their manufacture was based on a broadly available resource, evidenced by the large quantities that they have been found in, they established a degree of fairness across the group. Fairness was further enhanced in the collective idea spaces of latter lithic technologies that were employed by hominins that displayed a lesser degree of sexual dimorphism than their ancestors. Finally, stone tools permitted the user a greater degree of control over the vicissitudes of his or her environment.

The collective idea space of lithic technologies developed throughout the Paleolithic era in interesting ways across its various applications in the so-called stone tool 'industries'. The pace of this development may appear glacial in our estimation in hindsight. However, what these tools enabled a comparatively small group of individuals to achieve in terms of solidifying the basis of human progress on Earth really cannot be adequately compared with any other modern technology, no matter how transformative and scientifically advanced they seem to us now.

The reason for this is that the advent of lithic technology occurred on the basis of almost no prior knowledge. Rather, it initially occurred through the consolidation and realizability of core ideas in the various *individual* idea spaces that were constituents of an inchoate collective idea space, based closely on fundamental core principles alone.

Gradually, this consolidation and realizability began occurring on the basis of new ideas generated by associations *across* individual idea spaces in the collective idea space. Only toward the end of Paleolithic era were new ideas being generated in lithic technology through the association of *different* collective idea spaces altogether.

Due to the overwhelming association of stone tools with hunting that one may have in mind, especially those of the earliest stone tool industries in the Paleolithic era when stone tools had virtually no other use but the provision of food, we might be tempted to link the collective ideas spaces of hunting and lithic technology. However, this would not just be a somewhat tenuous, but possibly even an altogether incorrect working assumption. As a matter of fact, anthropologists have debated the relationship between the manufacture of stone tools, the bases for social organization, and the emergence of a language, and generally argue in favor of a thesis that they may have all emerged co-dependently.[5] Hunting, as a social activity that routinely relies on communication and coordination, was, therefore, quite likely *not* a feature at the initial stages of this evolutionary process (Krantz 1968).

It is perhaps worth a digression at this juncture to outline a few facts about the Paleolithic toolmaking or flintknapping industries.

At roughly the beginning of the Paleolithic more than three million years ago—that is to say, the Lower Paleolithic—the manufacture of stone tools was begun by early hominin species, chiefly, though not exclusively, Homo habilis. Soon this became an all-consuming enterprise, with stones being lugged from a significant distance to be reworked into tools; based on the location of its initial discovery in Tanzania, this first effort is now called the Olduwan industry. The technology used comprises a basic process of lithic reduction, whereby smaller flakes were struck off from a suitable stone using a spherical hammerstone to produce fractures with sharp edges. These could be large or small in size, and often the flakes that were struck off could themselves be used as tools. Archaeologists believe that these tools were not used for hunting, but rather for scavenging (i.e., scraping,

[5]See Gibson (1991), Tomasello et al. (2012) and Stout et al. (2008).

deboning, cutting flesh, and possibly smashing vegetable materials) (Hayden 2008).

The Olduwan industry lasted for a period that was a little short of a million years before Homo ergaster and, subsequently, Homo erectus, inherited the industry and proceeded to develop the technology into what is now called the Acheulean industry (Fagan 1994).

The improvement appears to be principally that the stone tools were bifacial and had forms that permitted them to be more versatile in their usage; generally, they were shaped on two sides into a pear-shaped stone using hammerstones or bones that caused the stone to crack through the application of pressure, yielding a tool with smoother surfaces on two sides and with relatively sharp edges. The most common types of tools this process generated were the handaxe and the cleaver. It appears to be far from settled in the literature whether these handaxes could have been used as projectiles in a hunt. Some experiments suggest that even from the outset of their development they may have been used for hunting; others suggest that the possibility is a remote one.[6] The fact is that the Acheulean industry, as initiated by Homo ergaster, does show a distinct technological development from the Olduwan industry.

So, regardless of whether they were hunters, this advancement alone is enough to motivate a consideration on whether they *thought* differently. Some researchers have argued that, since Acheulean tools represent symmetry and an enduring 'industrial process', perhaps its users not only had access to a higher and purposive cognitive functioning, but may have initiated social organization to a degree that enabled the process to be preserved through rudimentary channels of social learning (Holloway 2008).

There are some strong reasons to doubt this was the case, though, including the fact that there is debate over whether its earliest users could have even developed a basic language. Experimental evidence suggests that the stages involved in Acheulean toolmaking do indicate activity in the neural circuits of the brain that are associated with coordination between the visual and motor processes and in the

[6]See O'Brien (1981), McCall and Whittaker (2007), and Samson (2006).

development of language (Stout et al. 2008). What is significant is that the Acheulean tools were also used by Homo erectus. Variants of Homo erectus, from Africa to Asia, the Middle East, and Europe, all used aspects of this technology, but adapted it, often dramatically, to suit their own environments.

We know, for example, that Homo erectus did travel impressive distances across continents, including perhaps by sea as well, which suggests that they were also more socially organized than their predecessors had been. They also very likely did engage in hunting, at least in some regions, and possibly also cooking, given their indubitable mastery of fire.

The Acheulean industry lasted significantly longer than the Olduwan industry, and, therefore, holds the title for the most enduring industry in human prehistory. And it was at least midway during the Acheulean industry that, with the help of what was now the collective idea space of toolmaking, a different collective idea space emerged; art made its first appearance, suggesting that the new contexts also began to inspire Homo erectus in the creation of this new collective idea space (Toth and Schick 2015; Brahic 2014). However, just like it was with stone tools, representational art in sculptures and paintings took a long time in its development, and did not really become a relatively widespread pursuit till the Upper Paleolithic era.

The Acheulean industry gave way to the Mousterian industry a little more than 150,000 years ago, based on the Levallois technique. The technique involves flaking of pieces from all sides of a larger stone and then extracting a single piece from the center of this stone as a tool that has pre-finished, sharpened edges. It created very effective projectile points and was used by Neanderthals and Homo sapiens extensively in hunting.

Stone tools that were made using the Levallois technique suggest a significantly greater measure of purpose and foresight regarding the desired outcome than the Acheulean tools did; in other words, these were tools that were born from an involved 'design process', likely informed by collaborative thinking, or at least the sharing of ideas.

The Mousterian industry was replaced in the Upper Paleolithic by the Chatelperronian industry among Neanderthals some 45,000 years ago, which appears to have been Mousterian tools that were slightly enhanced, often just cosmetically, perhaps using cues from the tools made by neighboring groups of Homo sapiens. An example of a later Upper Paleolithic Homo sapien industry is the Aurignacian, which produced tools from bone as well as stone and emphasized use of the form of blades rather than flint-edged implements; a great variety of art and the first musical instrument, a flute, is attributed to the Aurignacian time period that began approximately 40,000 years ago.

This brief summary glosses over enough detail to make an anthropologist uncomfortable. It does not consider, for instance, the physical and environmental differences across the hominins that characterized the different industries, and neither does it consider the variety of other industries that began arising toward the end of the Paleolithic era and into the Copper Age. Besides, it may also give some of us reason for pause to draw any useful conclusions pertinent to a theory of ideas relevant for today from looking at our predecessors from that far back.

Still, our interest here in examining these industries is not to engage in armchair paleoanthropology, but to suggest the relevance of this case to the general nature of collective idea spaces. And, more specifically, what the progression of toolmaking industries can tell us about the foundational collective idea space that was relevant to a host of other collective idea spaces, such as hunting and art.

First, we might note that these industries were almost unimaginably long-lived and featured only an exceedingly gradual rate of increase in the transformations of the ideas that their collective idea spaces contained; for much of their existence, they seemed to have eschewed almost anything by way of innovation and creativity, remaining closely moored to an idea space mechanism based on a stable lithic technology.

Second, while we are looking at all the industries as collective idea spaces, the first attempts were notably *individual* pursuits. The technology of the earliest lithic industries represented a limited number of basic ideas forming a crisp mechanism that served to integrate the core principles of the foundational idea space that pertain to the survival of the individual. The mechanism was crisp across all individual idea spaces in the social group, with little role for fuzzier ideas that could have yielded any innovations.[7] For this reason, they only seem to have enabled rudimentary social cooperation across members; in this regard, it has been noted, for example, that imitation as a channel for the diffusion of technological know-how grew markedly over the Paleolithic, from virtually non-existent at its start to being a primary method of learning.[8]

From what we know it seems reasonable to deduce that the transformations of the Olduwan to the Acheulean industry were enabled by hominin species (Homo ergaster and, especially, Homo erectus) who permitted the consideration of fuzzier new ideas; the reason for this may be accounted for by the fact that they had access to a larger brain or because they travelled and adapted to new environments, and very likely as a result of a combination of these and other interrelated factors.

Our theory identifies awareness and the density of ideas as key to innovation or the generation of new realizable ideas. In both respects, the earlier toolmaking industries were severely hamstrung. In terms of awareness, individuals were overwhelmed by concerns pertaining to survival alone, but, and perhaps more importantly, in terms of the density of ideas, they only had access to an astoundingly sparse population. It has been estimated that till the latter part of the Upper Paleolithic the population of the world was not much more than 25,000 individuals (Huff et al. 2010). New environments and new contexts were, therefore, all the more crucial to the generation of new ideas.

Finally, as collective idea spaces, the toolmaking industries transformed from functioning only on the basis of theoretical creativity toward a mix of theoretical and praxis creativity. The Olduwan industry was severely limited by the cognitive abilities of the individual members of the Homo habilis species, and their inability to engage in an organized collaborative process. Successive Homo species became increasingly *more* collaborative, and their collective idea spaces began benefitting from changes to their contexts as members of their species began venturing into different areas.

---

[7]Social learning was conspicuously absent in the earliest lithic industries. See McNabb et al. (2004).

[8]On this issue, see Jelinek (1977) and Bar-Yosef (2002).

5.4 Learning to Throw

Given this background on the toolmaking industries, drawn directly from our own evolution, we now consider one of the core rules that we considered for our overarching collective idea space of sports: throwing.

Throwing a stone as a weapon in a hunt, especially as a strategy involving a team of hunters, would have offered distinct advantages to our predecessors in the Paleolithic, a fact that did not escape notice from Darwin himself, and has been remarked upon since by several other scholars.[9] The earliest subspecies of Homo sapiens, and possibly even Homo erectus in the late Lower Paleolithic and the Middle Paleolithic, may have hunted by means of throwing shaped stone projectiles. Certainly, the Acheulean industry yielded handaxes and cleavers that could have been lethal as weapons, especially in the hands of individuals who were, after all, devoted to not a great deal more than perfecting this task.

Yet, it is unknown with what motion our earliest ancestors threw these weapons, and, therefore, it can only be conjecture what force they managed to impart to their projectiles.

There is increasing evidence—albeit perhaps not enough to form firm basis for a consensus—that spears, perhaps even fire-hardened versions of them, may have been thrown in anger no earlier than approximately 300,000 years ago. The assumption, naturally, is that they would have been thrown with the traditional graceful overarm motion of a javelin thrower. However, cave paintings of elaborate hunting scenes, with individuals depicted in groups and holding spears aloft above their shoulders, only came well into the Upper Paleolithic, not much before 45,000 years ago.

Nevertheless, this does not detract from the fact that the art and skill involved in the process of throwing an object has been a product of human evolution, and it ought not to give much surprise that there is some evidence that the human arm evolved to favor being more effective at hurling projectiles. It is easy to imagine the evolutionary

[9] See Darlington (1975) and Isaac (1987) for a review.

advantage that perfecting the skill of aiming and throwing would have provided early hominins; the skillful and strategic launching of projectiles from a distance allowed them to contend with a harsh environment populated with more powerful predators and more robust prey, even when they lived in small groups. By using throwing as a substitute for sheer muscular brawn in overcoming an adversary in physical combat, or, more generally, by using throwing as a strategy to extend their reach, these early hunters brought a wider variety of animals within their ambit more safely.

As mentioned above, there is some research based on experimental evidence which indicates that the Acheulean handaxes made by Homo erectus in the late Lower Paleolithic era, and found in copious quantities in a variety of locations, may have been used as projectile weapons. However, it is quite plausible that even Homo habilis, who scavenged rather than actively hunted, may have used Olduwan stone tools to throw at predators or rivals in an effort to separate the latter from their kills.

The fact that primates that do throw objects in anger, frustration, or mirth almost invariably do so with an underarm motion stands to bolster the evolutionary argument. Constrained as we are now by the limited evidence, we cannot definitively conclude that Homo habilis could only throw underarm and that Homo erectus was first to throw with an overarm motion. But we are left with some tantalizing clues. Research has shown, for example, that primates that have a reliably developed an ability to throw objects show a significantly greater degree of connectivity in their brains between the primary motor cortex and the premotor cortex compared to primates who do not throw (Hopkins et al. 2012). This very likely suggests a greater degree of control over the muscles in the torso as well as a more developed set of muscles in the shoulder, which are both needed to launch a projectile with some force and precision using an overarm motion.

Experimental research suggests reasons for why it is that throwing with an overarm or sidearm motion considerably accelerates a projectile compared to an underarm motion (Venkadesan and Mahadevan 2017). While it may look incredibly effortless when done gracefully and effectively, throwing a projectile with considerable force, velocity, and accuracy is a fairly intricate physical and physiological accomplishment, requiring torque generated from the twisting of the upper torso, which must also sway from a lean back into a lean forward (Hong et al. 2001). Since primates have a center of gravity that is far higher in the body than do humans, they aren't able to master the challenge of throwing with quite as much finesse as can ordinary humans, let alone athletes, hunters, and warriors, who set about perfecting the skill purposefully.

Besides stone tools and spears, other objects may have been thrown as well, but, since they were far less amenable to standing the rigors of time, are proportionately underrepresented in the archaeological records. There is, for example, an extensive literature on throwing sticks, which are thought to have been in use from prehistoric times. Earlier examples of throwing sticks may have been very basic, but their availability and utility is not hard to imagine even for our Lower Paleolithic ancestors, given the somewhat circumstantial evidence that they are often employed by primates. Naturally, later variants of the throwing sticks became increasingly sophisticated. Curved rabbit sticks were used in Ancient Egypt to hunt birds; boomerangs, which are suspected to have been in use in Babylonia and southern India, apart from their usual association with the Australian aboriginal tribes, demonstrate an exceptional understanding of design and strategy in hunting based on throwing (Nies 1914). The rungu, which is still used by the Masai of East Africa, and the Irish shillelagh, which is now mostly of symbolic value, are two other examples of throwing sticks used in war and hunting.

The existence and timing of a hunting strategy that would have begun with the systematic lobbing of a stone weapon at a prey from a distance and then followed up with a ground assault comprising either thrusting weapons or heavier weapons to deliver blunt blows is a topic of lively debate in the archaeological scholarship.[10]

Besides suggestive of advances in lithic technology, such a hunting strategy is evidence of higher cognitive abilities, social organization, and coordination mechanisms by way of language and signals. The fact that the earliest forms of hunting would have combined aspects of

[10]See Packer and Ruttan (1988).

throwing and striking is arguably a strong basis for hunting as a primitive collective idea space at a larger degree of aggregation for the class of sports that employ similar ideas. As a matter of fact, if hunting by early humans is not *the* basis for the foundational collective idea space for sports, it is evident that it certainly drew upon the same core ideas that a foundational collective idea space for sports would have done.

Consider the emerging similarities in the evolution of hunting among hominins to the foundational core ideas of sports as we have defined them: an object was thrown, be it a handaxe or spear; an object, often the same as the one used for throwing, was used for hitting or striking; the preys were killed cooperatively by individuals working as a team, and the fact that a successful outcome at the end of the hunt yielded nourishment and resources made it a rather strong motivator. 'Winning', in other words, had much greater stakes with hunting.

That early humans reimagined the process of throwing and striking in a hunt into some sort of game while they were not hunting actively is obviously unknown. There is a compelling line of research which suggests that spheroid flintknapped stones were purposefully created in the Lower Paleolithic—at least under the Olduwan assemblage—and were probably used for pounding fibrous roots, bones, shells, and nuts. Through wear, these may have developed more completely spherical surfaces and may even have been prized for use as especially effective artifacts. Later, round stones were verifiably used in slings.

Whether such stones were rolled along the ground between a group of individuals, or even lobbed to a hunter who then struck it to accelerate it in the direction of a prey is obviously not known. The fact that a struck stone's velocity off the face of a stick used as a bat—it's 'exit velocity', as it were—is higher than even the hardest thrown stone is perhaps a fact that our earliest ancestors may have noticed, and perfected by practicing it as a pastime. That they used this method of launching a stone in a hunt is unknown, perhaps unknowable.

Nevertheless, that spherical stones were not also valued for their ability to be rolled along the ground seems unlikely; that they may then have been struck by the throwing sticks that they also had at hand is, rather frustratingly, only conjecture. What is perhaps true is that, when hunting was as crucial to survival as it was to our ancestors, it seems almost more unlikely to believe that the hunt was *not* practiced in times of relative 'leisure', or as a method to train the younger generation. Generally, though perhaps not entirely surprisingly, the discipline of archaeology is not bubbling with information on the topic of early Stone Age children and their amusements! Still, there is concrete evidence of games that were played by very primitive societies in Australia and Africa that involved boys throwing non-lethal mud balls at one another with the object of learning accuracy and evasion (McDaniel 1906).

5.5 Steppes to Make a Collective Idea Space

Archaeological research suggests that an interesting transformation may have taken place from the middle of the Paleolithic era to its end—one that also provides us with an arguably under-appreciated channel for the generation and exchange of ideas.

The change was inspired by the fact that the availability of megafauna—the mammoth being the poster child of this class of animals, though most others were somewhat smaller—began dwindling. The reason for this change in fortune has been attributed by scientists to a combination of changing climatic conditions, excessive hunting, excessive foraging, and other factors (Webster and Webster 1984). For our excursive interest, what is important is that this paucity of big game served to refocus the attention of our ancestors on much smaller animals that were relatively more abundant, principally on account of their greater rates of reproduction (Stiner et al. 1999).

There were three effects of this relative scarcity of the larger animals that are interesting in the context of our exploration.

First, it forced populations of Homo sapiens to not just migrate, but to also effectively occupy tracts of land that varied in their geographic extent based upon their carrying capacity for a sufficiently large number of food sources needed for sustenance. Their new ecological reality blurred the line between hunter and gatherer, or indeed between a hunter and a herder of potential species of prey, including fish. Now finding themselves more favorably matched against the smaller prey, it became possible for a larger fraction of the social group to become directly involved in the provisioning of food.

Second, tackling a greater variety of smaller game with significant variance in their predator-avoidance strategies, rather than the more challenging albeit larger animals that they hunted before, brought about a cultural change in the social groups inspired by the fact that the risks associated with gathering food were pooled across a somewhat larger set of individuals.

However, just as this new food source, spread over broadly unproductive territory, can be seen as an impetus for mobility over an expansive territory for some of our ancestors, a different resource can be seen as an anchor for others. The role that fishing and, generally, a subsistence based on aquatic resources played at around the same time is a fascinating topic in archaeology and anthropology (Erlandson 2001). The consensus seems to indicate that those social groups of hunter-gatherers that adopted an aquatically oriented lifestyle became more rooted to their locations than their counterparts who inhabited the vast interiors. They created more densely populated communities with richer cultures. Later, when farming of domesticated grains began to be seen as a real alternative to gathering at the end of the Upper Paleolithic, the ability of early societies to rely of mechanisms to pool the risks associated with variability across harvest seasons was thus not entirely a new innovation.

There are at least two general aspects that accentuate the relevance of this backdrop to our application of the theory of ideas.

First, it is worth assessing the transformative effects on the collective idea spaces of social groups. For almost the entire span of the Paleolithic era, collective idea spaces were largely a meaningless concept. This was not because individuals of early hominins did not live together. They verifiably did. Rather, the notion of a collective idea space had little value because the social groups they did form were riveted exclusively in the foundational core principles to such an extent that they may as well have been understood as aggregations of largely homogenous individual idea spaces. Innovation of new rules, for this reason, was virtually non-existent.

Later, beginning in the Middle Paleolithic, however, there was an incipient expedience for social groups to create more explicit methods for pooling the risks they faced. This is significant because, almost by definition, pooling the risks that a community faces is a function that a collaborative social organization facilitates. The various types of social hierarchical relationships that emerge are, to a large extent, determined by the manner in which the environmental contexts that a social group experiences create variegated conditions for their instantiation.

In this regard, it is worth noting that researchers who have carefully studied the evolution of the hominin brain suggest that the growth of the brain's size is only half the story; much of this growth had been achieved early in the evolution of hominins from Homo habilis to Homo erectus. In that significant evolutionary step, the increase in the size of the brain has been attributed to the fact that Homo erectus, unlike its predecessor, was a hunter, and travelled, planned, and persevered in the pursuit of a prey over several days; this explicit introduction of strategy may have been facilitated by the increased size of its brain.

In the latter stages of the Paleolithic, there was no remarkable growth in brain size. However, it was being gradually rewired toward becoming a more 'social brain'; sociality can be seen as a trait that was being selected for among hominin species (Dunbar and Shultz 2007).

It is hard to over-stress this point. The evolution of the brain was less about the needs for innovation and brilliant new ideas and more about learning to live in complex social groups; if it was the former, we would have seen a path of evolution in new ideas and new technologies that closely matched that of the evolution of hominins themselves rather than seeing almost all of that growth in the shortest span of time, and by Homo sapiens alone. In other words, the more it became possible for a collective idea space to be moored by its core ideas even as the number of individual idea spaces it contained was growing, the more it became possible for complex collaborative behavior to become the foundation for a staggering range of new ideas.

The second aspect from this evolutionary premise—one that served to enhance the fecundity of collective idea spaces—is facilitated by an interesting dynamic between social groups that were relatively stationary and those that were mobile. An expression of this form of social organization has existed in one form or another ever since then. A direct descendent from Paleolithic times, it is to be found in the contrast between the nomads and the settled communities that fringed their territories. The differences between the two are notably stark. They are exemplified in the wandering and pastoral lifestyle of the organized bands of nomads—crucially, the 'horsed' nomads—and barbarians. These nomads were indigenous to the steppes, deserts, and the mountainous regions located on the perimeters of several ancient settled societies situated on the coasts or on fertile river banks.

A great deal of interesting research suggests that these nomadic social groups interacted with the established settled societies on the fringes of their vast territories, acting as key agents of change in the process (Guzman 1988). Their superior abilities as horsed herders arose from a necessity to master the vagaries of the large expanses of their environment in order to sustain their growing herds. And they also possessed famously keen abilities as warriors, especially archers, which made them constant threats to settled societies. Yet, by most accounts, they usually assimilated well with the societies that they conquered, or, less frequently, that they came into contact with peacefully. The admixture of their leadership and strategic combat skills and the traditions of the settled societies led to initial periods of prosperity when the nomads vanquished the settled societies. Gradually, however, the transformation to becoming a settled and geographically bound society—concerned with governing, trade and feudal ownership—seems to have had the effect of diminishing their nomadic skill set, thereby converting their community to one that was again vulnerable to a new wave of attacks from other raiding tribes of nomads and barbarians.

Thus, the role of nomadic societies cannot be overstressed in any investigation on the processes by which innovations arose and were diffused across societies. There are boundless examples of such influences, including the Hyksos on Egypt in the seventeenth century BC; the Scythians on Eurasia in the ninth century BC; the Mongolian nomadic hordes since the third century AD, but especially the Pax Mongolica on Eurasia in the thirteenth century AD, and the Great Migration across Europe by the nomadic hordes in the first millennium AD.

The preceding observations suggest that the exigency for a dispersed society commenced in prehistory, and, over millennia, so well was this way of life honed that it was routinely used as the basis for nomadic warriors to organize and even conquer settled societies. The practice of guerrilla warfare and many of its associated strategies are derived directly from the nomadic hordes that made them their modus operandi.

All of this suggests that the nomadic societies routinely contended with and were successful at resolving the problems associated with collective action. The coordination of preferences and the management of expectations in outcomes that invariably produce winners and losers among a social group that was dispersed over extremely large territories are tasks that become decidedly more challenging to resolve as the size of a social group expands.

While the research on how this problem was addressed by the various nomadic societies appears to be far from settled, an essential mechanism seems to be that of a dynamic hierarchy. A dynamic hierarchy—as opposed to the more formal and stable one used in settled societies—permitted the flexibility that was needed to cater to the contextual differences across the regions that were held together in a loose coalition. The use of a system of intermarriage, tributes, taxes, and levies enabled fine-tuning the degree of centralization and control.[11]

The interaction between settled and nomadic societies provided both a medium of idea exchange across vast territories that separated settled societies, as well as, crucially, the introduction of specific ideas pertaining to their way of life into settled societies. The introduction of the composite bow is possibly the least disputable of these introductions, but the chariot, and many other aspects of horse-riding have also been attributed to them.

The World Nomad Games collects a range of historic nomadic games that are revived and played competitively by countries that would have been on the fringes of early Mongolian nomadic civilizations. Among these are several rather interesting games that combine aspects of hunting, horse-riding, and wrestling, including er enish, which is essentially wrestling *while* on horseback. While no direct evidence exists that it was the nomadic societies who were instrumental in the introduction of

[11]See Smith (1970) and May (2006).

polo to Persia and the Byzantine Empire in the sixth century BC all the way to China in the sixth century AD, the exceptional equestrian skills that they possessed make this a likely possibility (Liu 1985). The word polo is derived from a Tibetan language, and the variants played across the different regions on the fringes of the nomadic Mongolian societies had several similarities.

References

Bar-Yosef, O. (2002). The Upper Paleolithic Revolution. *Annual Review of Anthropology, 31*, 363–393.

Brahic, C. (2014, December 6). Etched in time. *New Scientist*, 2998.

Carter, J. (1985). Sport, War, and the Three Orders of Feudal Society: 700–1300. *Military Affairs, 49*(3), 132–139.

Dakhlia, S., & Pecorino, P. (2006). Rent-Seeking with Scarce Talent: A Model of Preemptive Hiring. *Public Choice, 129*(3/4), 475–486.

Darlington, P. (1975). Group Selection, Altruism, Reinforcement, and Throwing in Human Evolution. *Proceedings of the National Academy of Sciences of the United States of America, 72*(9), 3748–3752.

Dunbar, R., & Shultz, S. (2007). Evolution in the Social Brain. *Science, 317*(5843), 1344–1347.

Erlandson, J. (2001). The Archaeology of Aquatic Adaptations: Paradigms for a New Millennium. *Journal of Archaeological Research, 9*(4), 287–350.

Fagan, B. (1994). Timelines: Elusive Homo Erectus. *Archaeology, 47*(4), 14–67.

Gibson, K. (1991). Tools, Language and Intelligence: Evolutionary Implications. *Man, 26*(2), 255–264.

Guttmann, A. (1996). *Games and Empires*. New York: Columbia University Press.

Guzman, G. (1988). Were the Barbarians a Negative or Positive Factor in Ancient and Medieval History? *The Historian, 50*(4), 558–572.

Harris, H. (1963). Greek Javelin Throwing. *Greece & Rome, 10*(1), 26–36.

Hayden, B. (2008). What Were They Doing in the Oldowan? An Ethnoarchaeological Perspective on the Origins of Human Behavior. *Lithic Technology, 33*(2), 105–139.

Holloway, R. (2008). The Human Brain Evolving: A Personal Retrospective. *Annual Review of Anthropology, 37*, 1–19.

Hong, D., Cheung, T. K., & Roberts, E. M. (2001). A Three-Dimensional, Six-Segment Chain Analysis of Forceful Overarm Throwing. *Journal of Electromyography and Kinesiology, 11*(2), 95–112.

Hopkins, W., Russell, J., & Schaeffer, J. (2012). The Neural and Cognitive Correlates of Aimed Throwing in Chimpanzees: A Magnetic Resonance Image and Behavioural Study on a Unique Form of Social Tool Use. *Philosophical Transactions: Biological Sciences, 367*(1585), 37–47.

Huff, C. D., Xing, J., Rogers, A. R., Witherspoon, D., & Jorde, L. B. (2010). Mobile Elements Reveal Small Population Size in the Ancient Ancestors of Homo Sapiens. *PNAS, 107*(5), 2147–2152.

Isaac, B. (1987). Throwing and Human Evolution. *The African Archaeological Review, 5,* 3–17.

Jelinek, A. J. (1977). The Lower Paleolithic: Current Evidence and Interpretations. *Annual Review of Anthropology, 6,* 11–32.

Jessup, R. (2009). Transfer of High Domain Knowledge to a Similar Domain. *The American Journal of Psychology, 122*(1), 63–73.

Kaplan, H., & Hill, K. (1985). Food Sharing Among Ache Foragers: Tests of Explanatory Hypotheses. *Current Anthropology, 26,* 223–246.

Kilduff, G., Elfenbein, H., & Staw, B. (2010). The Psychology of Rivalry: A Relationally Dependent Analysis of Competition. *The Academy of Management Journal, 53*(5), 943–969.

Krantz, G. (1968). Brain Size and Hunting Ability in Earliest Man. *Current Anthropology, 9*(5), 450–451.

Liu, J. T. C. (1985). Polo and Cultural Change: From T'ang to Sung China. *Harvard Journal of Asiatic Studies, 45*(1), 203–224.

May, T. (2006). The Training of an Inner Asian Nomad Army in the Pre-modern Period. *The Journal of Military History, 70*(3), 617–635.

McCall, G., & Whittaker, J. (2007). Handaxes Still Don't Fly. *Lithic Technology, 32*(2), 195–202.

McDaniel, W. (1906). Some Passages Concerning Ball-Games. *Transactions and Proceedings of the American Philological Association, 37,* 121–134.

McNabb, J., Binyon, F., & Hazelwood, L. (2004). The Large Cutting Tools from the South African Acheulean and the Question of Social Traditions. *Current Anthropology, 45*(5), 653–677.

Nies, J. (1914). The Boomerang in Ancient Babylonia. *American Anthropologist, 16*(1), 26–32.

O'Brien, E. (1981). The Projectile Capabilities of an Acheulian Handaxe From Olorgesailie. *Current Anthropology, 22*(1), 76–79.

Packer, C., & Ruttan, L. (1988). The Evolution of Cooperative Hunting. *The American Naturalist, 132*(2), 159–198.

Pauketat, T. (2009). America's First Pastime. *Archaeology, 62*(5), 20–25.

Riess, S. (1994). From Pitch to Putt: Sport and Class in Anglo-American Sport. *Journal of Sport History, 21*(2), 138–184.

Samson, D. (2006). Stones of Contention: The Acheulean Handaxe Lethal Projectile Controversy. *Lithic Technology, 31*(2), 127–135.

Seippel, Ø. (2008). Sports in Civil Society: Networks, Social Capital and Influence. *European Sociological Review, 24*(1), 69–80.

Smith, J. (1970). Mongol and Nomadic Taxation. *Harvard Journal of Asiatic Studies, 30,* 46–85.

Speck, F. (1944). Catawba Games and Amusements. *Primitive Man, 17*(1/2), 19–28.

Stiner, M., Munro, N., Surovell, T., Tchernov, E., & Bar-Yosef, O. (1999). Paleolithic Population Growth Pulses Evidenced by Small Animal Exploitation. *Science, 283*(5399), 190–194.

Stout, D., Toth, N., Schick, K., & Chaminade, T. (2008). Neural Correlates of Early Stone Age Toolmaking: Technology, Language and Cognition in Human Evolution. *Philosophical Transactions: Biological Sciences, 363*(1499), 1939–1949.

Thorpe, I. (2003). Anthropology, Archaeology, and the Origin of Warfare. *World Archaeology, 35*(1), 145–165.

Tomasello, M., Melis, A., Tennie, C., Wyman, E., & Herrmann, E. (2012). Two Key Steps in the Evolution of Human Cooperation: The Interdependence Hypothesis. *Current Anthropology, 53*(6), 673–692.

Toth, N., & Schick, K. (2015). Evolution of Tool Use. In M. P. Muehlenbein (Ed.), *Basics in Human Evolution*. Amsterdam: Academic Press Elsevier.

Tremblay, V. (2009). Introduction: Economic Issues in Sports. *Review of Industrial Organization, 34*(1), 1–4.

Venkadesan, M., & Mahadevan, L. (2017). Optimal Strategies for Throwing Accurately. *Royal Society Open Science, 4,* 170136.

Webster, D., & Webster, G. (1984). Optimal Hunting and Pleistocene Extinction. *Human Ecology, 12*(3), 275–289.

6

The Ideas in Sports

6.1 The Case for Hunting

Let us now revisit the observation that the foundational collective idea space for a broad class of sports can be described with core rules that have close parallels with those that are associated with the collective idea space for hunting.

Our brief review of the origins and evolution of hunting suggests that the mechanisms of lithic technology and collaborative hunting were improved and developed over several hundreds of thousands of years.[1] It also suggests that hunting was a key basis for the migration of individuals and an early impetus for the transmission of ideas between environmental contexts; in this respect, it was perhaps not alone, but it was certainly a crucial component.

And, relatedly, it points to the importance of communication in the collective idea space of hunting as being a motivation for language. The idea of a hierarchy of fuzzy rules, central to our theory of ideas,

[1]More generally, the history of carnivory has been linked to the dynamics of human migrations and evolution. See Stiner (2002).

P. Goorha and J. Potts, *Creativity and Innovation*,
https://doi.org/10.1007/978-3-319-94884-3_6

provides some useful clues in both instances. Generally, the absence of a system of representing ideas—that is to say, the lack of a full-blown language—naturally delimits the degree to which a fuzzy core rule can be made crisper through the development of subsets representing any greater specificity.

Hunting assumed an overwhelmingly central role in the foundational worldviews that the earliest social groups of hominins subscribed to, and, as a result, it is not surprising that the core ideas of the collective idea space for hunting naturally drew upon the foundational core principles of safety, longevity, fairness, and control rather directly. For example, the ideas of manufacturing tools for the purposes of hunting and defense obviously served to enhance the social group's safety. That these tools were made in abundance, with very little alteration over a large span of time, suggests that they contributed toward the longevity of the worldviews of the social groups that employed them. However, on its own, lithic technology is a collective idea space that does not suggest itself as being at a larger degree of aggregation than hunting for a collective idea space spanning a range of sports. One reason for this is that the imperative to hunt was critical to the introduction of the core ideas pertaining to aiming and throwing projectiles, including stones, sticks, and spears, to those on lithic technology.

Arguably, successfully developing a collective idea space on hunting yielded core mechanisms that eventually helped replaced the core rules that pertained to scavenging in the worldviews of the Lower Paleolithic social groups of early hominins. This transition of early hominins from scavenger to hunter was not rapid. Not only did it come about after the largest increase in the size of the brain of a hominin species compared to its predecessor, but it also eventually enabled humans to engage in migrations that entailed adapting to vastly different environments.

And the transition also required a considerable amount of time because it depended on the ideas on lithic technology to first be associated with ideas that enabled perfecting the skills of throwing and striking, which themselves were made possible by appropriate physiological and cognitive changes in the hominin species as well as a radical alteration in the worldviews of their social groups. Further, hunting also required a degree of strategic planning and collaboration in teams that, perhaps for the first time, suggested a role for the creation of collective mechanisms that enhanced control and fairness in both the creation and distribution of rewards to encourage sustained participation by hunters and gatherers alike.

More importantly, however, this transition was perhaps also one of the first times in the history of human evolution that the role of prestige among individual members of the social group became especially relevant. We introduced the importance of prestige in an earlier chapter as an important mechanism by which the social group benefits from new rules that an individual develops and then shares with the group, thereby assisting the development of the entire collective idea space. The individual gains by accruing prestige—usually in competition with other members of the group with prestige—which can then be used in extracting benefits from the social group, such as obedience, protection, and respect.[2] With hunting, prestige is likely to have had a crucial role on several fronts, much of which is also common to predators that live in social groups. It encourages individuals to assume a disproportionately greater degree of risk in the pursuit of a shared benefit than they might otherwise have been be willing to take on for themselves, and it permits the social group to devise a coordinating mechanism with an individual leader, or even a hierarchy, that can marshal the efforts of others in tackling more complex tasks.

The association of the collective idea space of hunting to sports relies on the essential function that the ideas realized by the collective idea space of a sport serve for an individual's idea space. This can be seen in two interesting respects that are worth some consideration.

First, participating in a sport may serve the function of engaging in a simulation of core ideas. Ideas may well be entirely crisp enough for them to be considered core for a collective idea space, but they might only become core rules within an individual idea space once they have been learnt. There is a reason to believe that a sport can be seen as a

[2]In Butovskaya (2000), the author discusses the role of prestige in the course of human evolution in contrast to chimpanzees. A key point is that there are essential similarities across the two groups. In this context, see also Riches (1984).

natural learning mechanism—a simulation—for the core rules of hunting for individuals.

Studies in psychology suggest that we are born with an intuitive capacity to understand several aspects of Newtonian physics, and we then hone our predictive capacities through our experience with the real world (Hespos and vanMarle 2012). More than just routines of physical motion that pertain to the behavior of projectiles or biomechanics, it has been argued that sports are also especially adept at improving the predictive models that individuals rely on for other aspects that are foundational to their success (Battaglia et al. 2013). This includes developing an understanding of the behavior of individuals within teams through strategic collaboration and competition. Given the uniquely foundational relevance of hunting to survival for a preponderance of human existence, it can perhaps be argued that any activity that comprised ideas that were based on the core ideas of the collective idea space of hunting would have been of indispensable value to a social group if they could make the ideas crisper within the individual idea spaces of its members.

So, can the actions and rules of the class of sports we are considering, in this general sense, be seen as the basis for a simulation of hunting? Indeed, there are valid reasons for believing this to be the case.

Studies conducted on a range of animals have suggested that engaging in play effectively teaches individuals the concepts that are necessary and relevant to other social activities, of which hunting is the most prominent (Fein 1981). The fact that it is sometimes hard for an observer to distinguish between play fighting and more serious-minded battle suggests the close association of behaviors between the two activities. A large variety of animals have been observed to engage in variegated and complicated patterns of social play, especially play fighting, and studies have provided a range of reasons for such behavior, including physical and cognitive training, limiting aggression and providing vital information on how to engage with other members of a group (Cordoni 2009).

Group play among predators routinely includes simulated hunting techniques, including tripping and biting; solitary play has also been examined by researchers as a mode of play whereby the individual learns how to interact with inanimate objects and learn certain actions, such as twisting, leaping, rolling, and running. However, the overarching role of social play seems to be that of achieving behaviors that permit cooperative outcomes to be sustained.

While predators in the wild still retain hunting and play as natural behaviors, as the Paleolithic era gradually gave way to specialization and trade during the Neolithic revolution, the relevance of hunting was drastically dulled, even entirely removed, from the worldviews of more modern hominins. Sports then served the function of entertainment in settled societies, but they also supplanted the crucial relevance of hunting in the worldviews of social groups.

This suggests the second point worth noting: the advent of even partially 'organized' sports in recorded history is usually only found late into the Upper Paleolithic, well after somewhat larger and more settled social groups began forming communities. Once sports stopped serving the role of simulating the core ideas for hunting, they then became, more directly, among the broadest collective idea spaces that most individuals within a social group ever participated within; in effect, and more sweepingly, the collective idea space for sports then almost entirely replaced the need for participating in the collective idea space for hunting.

It would be misleading to say that sports were a perfect substitute for hunting in societies that no longer required most of its members to engage in hunting for its survival, or to even learn its core rules. Hunting remains a 'sport' in some countries even today. There may have been others; indeed, given that the collective idea space of hunting operated at a much larger degree of aggregation than that of sports, it would be surprising if there were *not* several others at lower levels of aggregation.

An equally appealing case could, for example, be made for dancing. The idea of a selfish herd captures the imperative felt by a group of prey to congregate, marking the resulting mass of prey appear more daunting to a predator (Hamilton 1971). For a group that contains individual members of a species of prey, moving rhythmically in unison makes the illusion of a single, larger creature yet more convincing to a potential predator. While this may broadly ally the core ideas of individual and organized movement in the collective idea spaces of dancing and hunting, the connection is perhaps more tenuous than it is for sports.

6.2 The Case for Herding

A sport that is based on some small object to strike—the ball—and some instrument to strike it with—the bat—comprises a label for a set of ideas that can have a number of interesting dimensions. Here, we are only considering the general nature of the core rule of striking or hitting that a sport may have borrowed from a foundational collective idea space that it had in common with hunting.

Consider bats. That simple clubs with a family resemblance to a baseball bat were one of the earliest weapons used by prehistoric man is a matter of more dispute than one would imagine. There is strong secondary evidence for their use as weapons or for ritualistic duels. However, wooden clubs simply aren't built to survive the rigors of time and, consequently, do not turn up with any regularity in the course of fieldwork that archaeologists undertake on the time scales that apply to early humans. On the other hand, there is much more overwhelming forensic evidence among animal fossils recovered at numerous excavation sites that indicates a kill having been made with stone arrowheads, delivered by means of a spear thrown by hand or an arrow launched from a bow. Naturally, this is not to say that clubs were not a weapon that was broadly available in the Stone Age, only that we are either faced with a truncated dataset by virtue of the greater longevity of stone as opposed to wood or that wooden bats and clubs may have been used sparingly since they were really a less ideal alternative in a hunter's arsenal.

In addition to hunting, there is more than a passing connection in the basic rules of the sports we are considering to the earliest practices of herding. While the practice of nomadic pastoralism is exemplified by many tribes in various parts of the world even today, there is historical evidence to suggest that the earliest forms of herding began in the middle Paleolithic era. And, of course, it has been practiced continuously ever since (Chang and Koster 1986). As a matter of fact, the occupation of herder has a history that far exceeds that of any one of the ancient civilizations, and certainly predates the occupation of a professional warrior.

The earliest herders came well before the early attempts at selective breeding of potential prey began during the Neolithic revolution in the

Fertile Crescent. As discussed in the previous chapter, these early herders were, in fact, hunters who had shifted their attention from hunting the depleted megafauna to the greater variety of smaller animals on land and water, and they engaged in a form a selective maintenance of animal populations by crude methods such as culling, the use of fire and by limiting the territorial ranges of their prey.

These early experiments with the occupation of herding provide us with several clues on the origins of core ideas for a foundational collective idea space that may help connect the history of sports with the ancient prehistory of humans. Our interest is in the conceptual origins of a broad class of sports, as well as its spread to different regions of the world; the nomadic basis, and the practice of transhumance that the job of herding entailed, suggests an interesting mechanism for the transmission and exchange of ideas.

A method for the diffusion and exchange of ideas in an otherwise very sparsely populated Earth during the Stone Age was transformative to the evolution of hominins and their societies. And, much before the practice of settled agriculture became a genuine impetus for the formation of constellated human societies based upon specialization in trades, the exchange of ideas was arguably facilitated by the nomadic civilizations that were based on hunting and herding, with their large territories and frequent intermingling, both peaceful and violent, with the settled communities of their time.

There is scant comparative and historical research on the topic of the lives of middle Paleolithic herders that one may use to examine the range and evolution of its practice globally, but whatever does exist points to the complexity of nomadic pastoralism as a social activity for a tribe or relatively small social group. Since nomadic 'shepherdry' requires significant coordination across a group of dispersed people, it necessitates the creation of a decision hierarchy among the herders charged with the organization, delegation and information-gathering responsibilities; that these were transformative practices for early humans to master seems undeniable.[3]

---

[3]Agrawal (1993) provides an interesting investigation for one such nomadic tribe.

The management of a team on a field of play as a core rule for the collective idea space of a large class of sports seems to owe much to the ideas on managing a field in warfare; the game of chess, for example, uses this core rule as its entire premise very explicitly. However, the organization of individuals on a field, especially in those sports that feature a safe haven, seems to be an idea that has its roots in herding and hunting before warfare became a routine and necessary part of social life.

The evidence of herding leaving its mark on sports is often largely tangential, and far more recent, by which time herding was an entirely established and separate occupation from its roots in hunting. As an example, consider one of the most intriguing sources for field hockey's ancestry, which comes from Beni Hasan in Egypt. There, at the tomb site of a local governor of some importance in his time, dated to approximately 2000BC, is a depiction of two men. Each holds a stick with the slightest of crooks at the bottom end as they stand facing one another. They are bent at the waist as their sticks are in contact with something that looks either like a large ball or a small hoop.

What is interesting is that there was already a staggering variety of staffs and sticks in circulation by that time, many with pronounced hooks at one end, and some with the definite purpose of assisting shepherds with their work; moreover, the dual use of such staffs for leisure and for shepherdry is not hard to imagine (Fischer 1978). The impressive list of functions for the crook at the end of a shepherd's staff includes: useful in guiding animals by their necks; fending off predators; striking snakes on the ground; helping with walking on uneven terrain; and using it to hook high branches with in order to pull them lower for their herds. Arguably, therefore, a long stick with a crook at one end is even better suited for shepherdry than it now is for field hockey.

Besides the shepherd's staff with a crook, other forms of sticks were very likely part of the herder's toolkit. For example, the Queen Mary Psalter, compiled at the beginning of the fourteenth century, contains an impressive illustration depicting the practice of pannage, which involves permitting domesticated hogs to feed on acorns in a forest, often agisted by the forest's owner. In the vivid painting, two men are shown holding clubs that bear more than a passing resemblance to a baseball bat. One is swinging with a hammer grip at a tree that is dropping its acorns to the ground, where a grateful pig devours them. The other is swinging expansively, one-handed, at another tree with an acorn in midair.[4]

The use of a shepherd's staff as a bat has also been suggested in relation to the earliest forms of cricket. In what seems like rather rapid evolution for a game with a directly verifiable history that began in the sixteenth century, over the course of just 50 years in the eighteenth century the shape of the cricket bat morphed from what looked like something closely resembling a hockey stick to something that looked much closer to a baseball bat, and only then into a bat that resembles its modern counterpart.

The earliest versions of lacrosse sticks used by some native American tribes also appear to have be modeled on a shepherd's staff with netting spanning the end where the crook exists. Whether the Nordic game of knattleikr, which appeared to have more than a passing resemblance to baseball, used a bat that resembled a staff is not known with certainty; the mention of the bat being thrown at opponents and for hitting the ball considerable distances makes it hard to imagine how a lacrosse stick would have served the purpose well. However, the presence of Norse travelers in the Vinland region several centuries before Columbus suggests the tantalizing possibility that there may have been some exchange of ideas in both directions (Thurber 2015).

6.3 Some Thoughts on the Ideas in Sports

6.3.1 Throwing and Striking

Let us now resume our consideration of some sports that involve a few of the key core ideas in their overarching collective idea space.

Our discussion above on the evolution of the skill of throwing among hominids suggests that it is very likely that learning how to throw

[4]The Psalter can be viewed in full online at the British Library's website for the Catalogue of Illuminated Manuscripts. The scene depicting pannage is from a calendar page in November.

comprised several stages. The earliest of stages, arguably more than a million years ago, may have involved nothing more than barely rolling a spheroid object along the ground, perhaps to get it to its destination faster.

That even the earliest hominids had access to balls of some description is rather likely, not just because spheroidal stones form as a matter of course from the regular use of hammerstones, but also because they occur naturally as well; there is an interesting literature in geology on 'armored mud balls', for example, which are found all over the world, and that are formed by a process of the accretion of clay, pebbles, and sediment around a core through the action of water (Bell 1940).

Rolling round objects may have then led to the ideas of flinging the object over obstacles. However, from those initial ideas on rolling a ball to being able to throw one with an overarm motion, and to do so with a repeatable motion that gives speed, accuracy, and power, were clearly evolutionary leaps that were not made easily, and never made by our primate cousins; indeed, there is reason to believe that they may have been absent among the earliest hominins altogether.

Games that involve rolling a ball along the ground may, therefore, have been one of the earliest 'sports', though there is obviously no record of this or any other form of organized ball game having been played before the Neolithic age. Today, several games still exist that involve rolling a ball along the ground, though they all involve the skill of aiming at an object as well. Among these is the game of lawn bowls, which has a long lineage within the larger category of 'boules games' that include several similar games that are played on lawns, streets, rural tracks, and improvised surfaces. They involve aiming, throwing and rolling a ball, and scoring, though the aspect of a team that is actively fielding is not a feature common to most.

Bowling, more generally, has an ancestry that appears to comfortably exceed any other ball game (i.e., if one does not count catching a ball as a game), with evidence of it being prevalent in Egypt as early as 3200BC, but also in Rome before the third-century AD and in Germany as a religious rite at roughly the same time. Generally, there appears to be a variety of balls in use in early Greek and Roman civilizations, including those filled with air and feathers, and one even made of glass, though there is no ready evidence of the use of a game employing a bat that was used to strike the ball with. That a game did exist, called trigon, that involved using the hand to bat away balls from (or perhaps at) an opponent suggests, however, that it would not be surprising if a bat of some type may have been used.

Just as there are games that involve bowling without the aspect of striking with a stick, there are, of course, a large variety of games that seemed to have introduced striking with a stick to, essentially, a bowling game, almost as an afterthought.

Among these, we may count kolf, the most direct ancestor to golf, which is a thirteenth-century invention from the Netherlands that was initially played as a team game that required reaching a target. Jeu de maille is a fifteenth-century French game played with a mallet and was a cousin to jeu de paume, which was likely the ancestor of tennis, initially played indoors by the nobility with the open palm of the hand standing in for what is now the tennis racquet. Trucco is a seventeenth-century Italian game played with wooden balls and long cues with short paddles at the end and was played on lawns; the better-known game of croquet is only a mid-nineteenth-century English invention.

In speaking about sports that introduced a method to strike a ball to a game that resembled lawn bowling, we should arguably consider cricket. Prior to the mid-eighteenth century, bowling in cricket was hardly dissimilar to what it looks like in bowling or lawn bowls. The ball, in other words, was not pitched but rolled along the ground. The earliest painted depiction of cricket, dated 1743, shows this type of bowling in progress, or something very close to it. At least one history of cricket describes a very early version of the wicket to be very different to its modern reincarnation: it was barely a foot high, two feet wide, and lacked a middle stump altogether, since the object was to pass the ball 'through the gate' to effect a dismissal for the batsman. The evolution from underarm to round-arm bowling in cricket began in the second half of the eighteenth century, when a cricketer, Tom Walker, determined that round-arm bowling—which is a method of delivering the ball to the batter by launching it with the arm held further away from the body rather than close to it, as it would be with either underarm or overarm bowling—permitted the bowler to introduce some useful variations in his repertoire, as well as exploit the conditions of the surface better (Altham 1962). This fact makes it less likely that cricket owes its origins entirely to stoolball, as is commonly supposed, since all evidence for stoolball suggests that the ball was pitched and was swatted away with a bat that looked much like a table tennis racquet with a long handle.

What we are, therefore, left with are tantalizing clues that suggest that the advent of bat and ball sports may well have been in the earliest pages of the unrecorded history for humans. Recorded history alone clearly suggests—as it incontrovertibly does for a range of sports—that playing certain sports were seen as legitimate ways to prepare for war; it seems very likely that certain types of games would have also been seen as a routine way to prepare for a hunt ever since hunting became a way of life for our early ancestors. Aspects of the hunt would have been replicated in games and then endured as a primary source of entertainment well after the need for hunting abated.

It has been established, for example, that the ideal distance for a projectile thrown by hand for the purpose of hunting large animals is between 10 and 20 m, since, at this distance, a specific anatomical region on the prey's body can still be targeted accurately (Isaac 1987). That this translates rather well with the distances for pitching and bowling in baseball and cricket can, of course, be seen as entirely coincidental, but the prospect that the distance represents a zone to be able to ideally engage or target an adversary is less controversial.

6.3.2 Prosociality

As we have argued, the idea of prestige as an important social value has roots in hunter-gatherer and nomadic pastoral societies. Prosociality is simply seen as the behavior that is congruent with a society that values prestige. Individualism, by contrast, is logically a myopic strategy. Our earliest ancestors would have discovered that an individual's skill may not always directly correlate with hunting and foraging success; experience, for example, may be just as valuable, even in the absence of physical skill. As a result, providing longer-term prestige may have been emerged as substitute mechanism to directly monitoring short-term physical contributions to the group alone.[5]

Since the earliest societies depended on harmonious communal living, actions that encouraged prosocial behavior, were likely to have become an essential feature of the social traditions of successful social groups. It is obvious then that a norm on prosocial behavior draws directly on the core principles of the foundational collective idea space: it promotes interdependence among the members of the group in the face of uncertainty in the satisfaction of the essential needs of individuals within the group. For hunter-gatherer societies, these needs related to food supply and security, making it all the more essential that the members of the groups be encouraged to adopt such behaviors.

It is, of course, not known precisely what methods may have existed among these earliest of human communities that resulted in the enhancement of prosocial behavior, though the most likely candidates are very likely to have involved a mechanism for rewarding individuals over the course of their lives, with prestige or opprobrium.[6,7] Later, however, other activities may have continued to build on the idea of encouraging kinship among the group, including games and dances used as 'team-building exercises'. Generally, a norm of prosocial behavior would have encouraged cooperation across a range of activities and avoided reifying an aggressive dominance hierarchy that was unsympathetic to contexts beyond hunting and, importantly, antithetical to the core principles. Prosociality would also have been strongly reinforced by an affnity to a shared identity of the group that provided the prestige and recognized it.

What can be adduced, more generally, is that a set of rules may have incipiently emerged as the need grew stronger in the first diversified

---

[5]See Hill and Kintigh (2009) for an interesting review on this topic.

[6]Anthropologists have usually tended to confirm the evolutionary advantages that one would expect prestige to confer by looking at a range of contemporary tribal societies. The rewards include better access to mates of higher quality and, through rewards, higher chances for the success of offspring. See von Rueden et al. (2011) as an example.

[7]Game-theoretic models routinely show cooperation with the group as an evolutionary stable strategy. In examining alternate theories for the evolution of hunting, for example, Packer and Ruttan (1988) suggest that the most viable explanation for cooperation is 'gregariousness'.

societies of early humans to develop a system for keeping score of prestige and opprobrium that was generally acceptable. Within these rules, we are likely to find at least part of the motivation for organized social institutions based on the requirement of relinquishing control and power to an established set of leaders, such as those pertaining to religion and government.

The first evidence of religion among hominins dates back to the Middle Paleolithic, with evidence of burial sites and several hunter cults based on using the bones of bears as mythical symbols (Narr and Auer 1964; Germonpré and Hämäläinen 2007). Among these institutions, one might find the motivation for the advent of organized games as rituals. It is an established fact that sports were routinely organized as an offering to a deity or for the entertainment of a potentate. Indeed, there are examples of several games that explicitly introduced motifs and associations to the practices of worship or allegiance to religious authorities in their play. There are, for instance, vivid depiction of games that were played in ancient Egypt for the entertainment of the pharaohs (Wilson 1931). In India, the game of gulli danda is mentioned in the *Mahabharata*, and is, therefore, possibly the oldest game of its type, with a recorded history that stretches back over approximately 3000 years (Menon 2006). The game, still played in some parts around the world in addition to India, involves a thin wooden bat, which is used to strike and launch a wooden peg, then struck again. It also features the idea of bases and fielding. The Mesoamerican ball game, or ulama, dated to at least 2500BC, had strong associations with Mayan worship and sacrifice rituals. And, as mentioned above, the French game of la soule began outside the churchyards of the towns that participated in the game, often in teams of hundreds.

6.3.3 The Talent

The desire to seek prestige by prosocial individuals has the natural side-effect of making talent a desirable and valuable social good.

Yet, in the context of a group, the relative effects of talent across individuals are not linear; in other words, being even *slightly* more talented can often make an athlete several times more effective at the sport than her closest rival, making her replacement only a weak substitute to herself. A plausible reason for this is that talent in sports emanates from a degree of confluence in a specific set of skills that enables an individual to outperform the average rival she encounters and assists her, or her team, in winning a contest.

Relative talent between athletes follows a power law, which is to say that the ranking of some metric of the talent that an individual athlete possesses varies with a proportional relative change in another. This other variable can, of course, be the remunerations that athletes receive. The generous incomes of some professional athletes in sports are even more eye-catching when contrasted against the stratospheric incomes of the few elite athletes among them. Admittedly, this metric is less than reliable in a number of circumstances. It is, for example, easily influenced by regulatory hindrances or by the popularity of the sport in local markets. Besides, it cannot reliably be used across time periods in comparing the relative talent of athletes through the history of a sport. Having said that, it does generate some interesting research. For example, Pantuosco and Stone (2010) estimate that Babe Ruth would have comfortably received something in the vicinity of fifty million dollars were he playing in the league in 2005. In salary alone.

This nonlinear behavior between talent and its spoils notwithstanding, a second feature of talent is, perhaps, also equally inarguable: talent enables the accumulation of prestige commensurate with the size of the collective idea space of the pursuit.

In sports, this size of the collective idea space is proxied rather well by the size of the market. Prestige enables higher salaries and the adoration of the masses, the latter obviously facilitating the former, making the 'spoils' of prestige a natural function of the size of sports market; sports that have a smaller following permit a talented individual to generate more limited prestige within the context of the larger social group. The bigger the sport, in other words, the bigger the superstar (Rosen 1981).

In examining these features of talent, it is worthwhile stepping back and considering the role of talent from the perspective of our theory.

Taken tabula rasa, beginning with a foundational collective idea space, the distribution of control is symmetric across individuals. With this premise, as we have seen above, prosocial behavior then creates an environment within which individuals seek prestige as currency to effect change and create collective idea spaces at lower levels of aggregation. In a Paleolithic social group the most talented hunter naturally had the ability to generate a great deal more prestige than the most talented at just about anything else. Consequently, we should expect that it would be the most talented hunter who would have also been most likely to be the leader of the group in the group's other endeavors.

Within sports, for example, prestige accumulated on the basis of talent can provide the ability to make an impact on the fundamental nature of a sport. Rare talents often have the ability, directly or indirectly, to have outsized impacts on the collective idea space that their sports represent. Examples of changes in rules of the sport inspired by the outstanding talent of a player are numerous. Among them are the change from 12 to 16 feet for the foul lane in basketball due to Wilt Chamberlain, the lowering of the pitching mound in baseball due to Bob Gibson, and the introduction of leg theory to counter Don Bradman in cricket.

Using our theory, we ought to be able to assess whether talent is discernible in the individual idea spaces of the talented athletes in terms of ideas that an individual is aware.

Our observation in this regard is simple: talent can be seen in precisely the same manner as creativity. Specifically, what we perceive as exceptional ability in an individual can be based on either theoretical creativity or praxis creativity. When talent relies on theoretical creativity, the athlete innovates new ideas. When it relies on praxis creativity, the athlete innovates new perspectives on an extant idea and employs mechanisms in consolidating aspects. Since the latter method emphasizes consolidation in the collective idea space, it is more likely to yield public acclaim.

The difference is worth considering, since the two forms of creativity may sometimes be separated across individuals for the same idea in the collective idea space of a sport, and at other times embodied within a single person. The English cricketer, Bernard Bosanquet, for example, was a pace bowler who invented the googly delivery at the turn of the nineteenth century. Yet, he does not feature as an icon of the sport since, despite a few moments of sheer genius, he remained an unreliable bowler of this own invention. On the other hand, Shane Warne, an Australian leg-break bowler, is widely considered to be among the greatest bowlers of all time; his repertoire included the googly, but it was his skill in producing each delivery with uncanny accuracy that nourished his reputation as an outstanding bowler. In contrast, Dick Fosbury's technique for attempting the high jump—involving jumping over the bar in a supine position—required going against the advice of his coaches and the establishment in the early 1960s. Consolidation of the idea was not a consideration to him, and, as such, his was a new idea developed through theoretical creativity. However, his reward, in the form of a gold medal at the 1968 Olympics, only came after a great deal of hard work and practice; he had invented his new approach while still in high school.

In sports, especially team sports, the collective idea space represents several ideas that might be separated or conjoined in the course of what we observe in a few minutes of a match, let alone in an entire season or a career. The demarcation, therefore, between what is theoretical creativity and what is praxis creativity is often not clearly discernible. Indeed, we should expect this to be the case when new ideas also require the use of feasible new mechanisms so that they might be made crisper for others involved the sport, be they regulators or fellow athletes.

How a pitcher or bowler employs his repertoire of deliveries in baseball and cricket makes for an interesting example. Certain deliveries, such as the knuckleball in baseball or the carrom ball in cricket, are developed and honed largely by the individual athletes themselves, albeit they often rely on whatever morsel of wisdom they can glean from past practitioners. On the other hand, a talented pitcher of fastball variations or a bowler of express pace bowling relies on largely consolidated ideas within the sport's collective idea space; their talent is founded upon praxis creativity. The talent of such pitchers and bowlers often comes from being able to bank on their abilities to refine the ideas of a pitch or delivery to a highly crisp degree, be able to employ an effective repertoire, and develop insight on their rivals' weaknesses. Reverse swing bowling, where a cricket ball swings in the opposite direction to what its path would ordinarily be when the ball is newer, is an example of a delivery that cannot be employed by a bowler with much effectiveness unless all members of his team understand their role in curating the condition of the ball over the course of a match.

While our theory suggests that talent can usefully be examined as a composite of ideas that the athlete's idea space represents in relation to the sport's collective idea space, an obvious objection is that the physical gifts of an accomplished athlete can hardly be seen as secondary to his or her set of ideas.

The general consensus, however, appears to be that the narrative of the lazy superstar athlete who simply happened to have won the genetic lottery is a gross simplification, if not altogether a myth. Elite ability, it appears, comes routinely from dedication based on effortful cognitive functioning as well as a desire and ability to leverage certain genetic advantages and natural predispositions. It is worth underscoring that, just as superior physical abilities are not always observable—even entirely absent to the casual observer of a new sport—so, too, are an athlete's cognitive gifts. Besides, physical traits are often only a starting point; they contribute as a mechanism in realizing ideas within the idea space, but are rarely as 'ideas' in and of themselves.

There is indeed convincing evidence that suggests a key role is played by the uniqueness of an athlete's cognitive processes in the determination of her eventual success at a sport. An example of this, as outlined in Chapter 2, pertained to the differences between the abilities of a batter.

A less accomplished batter in baseball or cricket has to process a larger quantity of information in short-term memory alone, since he does not have access to the deep reservoir of pertinent information a more accomplished batter might have developed. The elite batter stows away a great deal of information in his long-term memory, rather like a personal portable reference library that he can rely on in any situation of the game in which he finds himself. He simply retrieves the information he requires from his long-term memory and introduces it to his working memory, thereby reducing the overall strain on his short-term memory. The fact that even elite batters sometimes struggle in contextual conditions that are unfamiliar to them suggests the validity of this simple model.

The aptitude test known as the Wonderlic, which is taken by potential prospects for the National Football League, provides another rather good example. While there are some systematic differences in the scores achieved by players at various positions of the game, there does not seem to be a strong indication that a high score is particularly necessary for a successful career, let alone necessary to becoming an elite football player.[8] Why then is the test administered to prospective players?

One rationale is simply that it helps a coach in the establishment of a benchmark, and provides some common measure on the ability of the player to take instruction from the coaching staff and from his peers adequately and effectively. If the team's playbook is a proxy for the collective idea space, a very low or very high score on a general aptitude test is only relevant if the player's own individual idea space does not have an idea balance with the core of the collective idea space already. This is rarely the case, since football prospects for the NFL have usually played many years of football prior to entering the draft. When they come to the NFL having never played before, naturally their exceptional physical ability is a crucial attraction. However, they then also tend to score higher than average on the Wonderlic for their positions. Stephen Neal, who made the switch from wrestling to a successful guard for the New England Patriots, is an interesting case in this regard.[9]

6.3.4 Teams

Cooperation is immanent in the core principles of a foundational collective idea space, and we have argued that the core rules for a class of sports emerged from that base through a rather expansive hierarchy of fuzzy rules that includes hunting and herding, though very possibly other collective idea spaces such as war played interesting roles as well.

---

[8]See Fry (Undated) and Deary and Der (2005) on this topic.

[9]As an aside, a different explanation may also exist for the continued use of the Wonderlic, despite claims that it isn't an accurate measure. Research by psychologists has established that an inverse relationship exists between IQ and mortality, which is significantly dampened when reaction time is controlled for (Leite 2009). Indeed, the Wonderlic scores for positions that do require higher reaction times—the quarterback and the blind-side tackle—are, very generally, somewhat higher. However, research also shows that the inspection time required for a task is also correlated with a faster reaction time, and, further, that it is correlated differently to reaction time than is IQ.

The rules for talent and prestige played a mutually supportive role in the creation of collective idea spaces for a variety of sports at lower levels of aggregation. Among such collective spaces, we count those of several sports that thrive on the successful formation of a team of individuals, selected from within the context of a larger group, who can work together effectively.

While this process of selection may appear to be based on a choice made on some required skill alone, it can be argued that this skill, in fact, stands as a visible signal for another latent characteristic that is even more strongly correlated with the core rule of the sport's collective idea space.

The characteristic is that of 'shared intentionality', and it is a quality that is at the heart of an evolutionary mechanism that enables sustaining collaborative behavior among the members of a team in pursuit of a common objective. It is worth noting that shared intentionality is different from prosocial behavior. In contrast with prosociality, it refers to the specific trait of having a common objective through a reduction in the idea imbalance between a group of individuals engaged in a pursuit relevant to shared subspaces.

And in this respect, it is arguably even *more* important than skill. This mechanism is able to explain why cooperative behaviors can be exceptionally stable in groups that may otherwise seem too large or diverse to be in a position to rely on prosociality alone, and yet are rarely large enough to be in a position to rest on a purely cultural explanation as the basis for what motivates broadly cooperative behavior.

This can be seen in a somewhat different way. In a smaller team, with well-defined mechanisms to achieve a common objective, and one that comprises paid professionals whose skill levels are well understood, team collaboration becomes relatively easy to sustain on the basis of making the marginal wage for each member as close to her marginal product as possible. When the objectives of the team begin to multiply, however, wage setting rapidly begins to lose meaning, since, not only does the prospect that the marginal values of the individuals will vary by each objective begin to arise, but also because each objective might be weighted differently by different members of the team.

With the shared intentionality of hunting or scavenging big game, early humans devised strategies for collaborating in groups that were able to circumvent the key problems regarding the individual's desire to free ride on the efforts of others and in ensuring that each individual received an equitable share in the spoils of the group's efforts. However, hierarchies may have developed even among the earliest hunters on the basis of encouraging sustained cooperation by the relatively more skillful by means of a larger share of the prize, in addition to greater levels of prestige.

Interestingly, there is strong evidence that this form of interdependent collaborative behavior is, to a high degree, uniquely human, and not generally observed in other primates. And that it is arguably both ontogenetic as well as phylogenetic, which is to say that the potential for collaborative behavior is something we develop from birth and that it is unique to our species (Fein 1981). Thus, it is worth examining how a baseline of collaboration and shared intentionality might emerge in a group of individuals even in the absence of encouragement by way of asymmetric shares of the prize or with different levels of social prestige.

It is possible to see how, in the context of a larger group, individuals may develop a shared intentionality by considering a simple setup with members who are dominant, equal or weak in terms of their skill at some task.

The collaborator's dilemma pertains to any general situation where a minimum of two individuals must collaborate to achieve some objective; once the objective has been achieved, a dominant collaborator may find the option of offering nothing to her less significant and weaker counterpart appealing. In the setting of a team or large group that features dominant and weak members interacting in pairs, if most dominant members feel compelled to act in an ungenerous manner toward their weaker counterparts, weak members (as well as any generous dominant collaborators) find themselves at a disadvantage. It isn't hard to see that such 'collaboration' does not give the overall group much of a chance at surviving unchanged; weaker members of the group may decide to leave or revolt against the injustice.

A possible solution that salvages a stable collaborative outcome emerges when the members are strategically and structurally compelled to interact in a manner that requires them to interact with a different set of individuals periodically, who may be weaker, equal in ability or more dominant than them.[10] A range of team sports, of course, do this as a matter of course. Why this should result in collaboration being a distinct possibility should be evident from the fact that a dominant individual cannot rely on her dominance forever, and, consequently, all members in the group more willingly 'learn' the behavior of collaborative engagement over the objective of the task at hand. In other words, they learn to have shared intentionality.

This example assumes that the real skill or talent level of an individual is unknown; in a group large enough where the probability of meeting the same individual repeatedly is small, or where the membership or access to a team is open, and new members with new levels of skill may enter at any moment, such an assumption may be safer to make.

All of this suggests that collaboration in humans quite possibly has a primitive basis in the core principles. It is reinforced at the level of the individual by self-interest, itself also bolstered by reciprocity among peers and concern for relations. And it is supported at the level of the society by institutions that embody, enshrine and protect those norms and rules that favor cooperative and altruistic behavior to aid its development and preservation.

References

Agrawal, A. (1993). Mobility and Cooperation Among Nomadic Shepherds: The Case of the Raikas. *Human Ecology, 21*(3), 261–279.

Altham, H. S. (1962). *A History of Cricket* (Vol. I). London: George Allen & Unwin.

Battaglia, P. W., Hamrick, J. B., & Tenenbaum, J. B. (2013). Simulation as an Engine of Physical Scene Understanding. *Proceedings of the National Academy of Sciences of the United States of America, 110*(45), 18327–18332.

Bell, H. (1940). Armored Mud Balls: Their Origin, Properties, and Role in Sedimentation. *The Journal of Geology, 48*(1), 1–31.

[10]See, for instance, Killingback and Studer (2001) and Tarnita et al. (2009).

Butovskaya, M. (2000). The Evolution of Human Behaviour: The Relationship Between the Biological and the Social. *Anthropologie, 38*(2), 169–180.

Chang, C., & Koster, H. A. (1986). Beyond Bones: Toward an Archaeology of Pastoralism. *Advances in Archaeological Method and Theory, 9*, 97–148.

Cordoni, G. (2009). Social Play in Captive Wolves (Canis Lupus): Not Only an Immature Affair. *Behaviour, 146*(10), 1363–1385.

Deary, I., & Der, G. (2005). Reaction Time Explains IQ's Association with Death. *Psychological Science, 16*(1), 64–69.

Fein, G. (1981). Pretend Play in Childhood: An Integrative Review. *Child Development, 52*(4), 1095–1118.

Fischer, H. (1978). Notes on Sticks and Staves in Ancient Egypt. *Metropolitan Museum Journal, 13*, 5–32.

Fry, B. (Undated). *Brains on the Line*. http://benfry.com/writing/archives/147. Accessed on May 4, 2018.

Germonpré, M., & Hämäläinen, R. (2007). Fossil Bear Bones in the Belgian Upper Paleolithic: The Possibility of a Proto Bear-Ceremonialism. *Arctic Anthropology, 44*(2), 1–30.

Hamilton, W. D. (1971). Geometry for the Selfish Herd. *Journal of Theoretical Biology, 31*, 295–311.

Hespos, S. J., & vanMarle, K. (2012). Physics for Infants: Characterizing the Origins of Knowledge About Objects, Substances, and Number. *Wires Cognitive Science, 3*(1), 19–27.

Hill, K., & Kintigh, K. (2009). Can Anthropologists Distinguish Good and Poor Hunters? Implications for Hunting Hypotheses, Sharing Conventions, and Cultural Transmission. *Current Anthropology, 50*(3), 369–378.

Isaac, B. (1987). Throwing and human evolution. *The AfricanArchaeological Review, 5*(1), 3–17.

Killingback, T., & Studer, E. 2001. Spatial Ultimatum Games, Collabo-Rations and the Evolution of Fairness. *Proceedings: Biological Sciences, 268*(1478): 1797–1801.

Leite, F. (2009). Should IQ, Perceptual Speed, or Both Be Used to Explain Response Time? *The American Journal of Psychology, 122*(4), 517–526.

Menon, R. 2006. *The Mahabharata: A Modern Rendering*. Bloomington: iUniverse.

Narr, K., & Auer, N. (1964). Approaches to the Religion of Early Paleolithic Man. *History of Religions, 4*(1), 1–22.

Packer, C., & Ruttan, L. (1988). The Evolution of Cooperative Hunting. *The American Naturalist, 132*(2), 159–198.

Pantuosco, L., & Stone, G. (2010). Babe Ruth as a Free Agent: What the Old-Time Greats Would Earn in Today's Labor Market for Baseball Players. *The American Economist, 55*(2), 154–161.

Riches, D. (1984). Hunting, Herding and Potlatching: Towards a Sociological Account of Prestige. *Man, 19*(2), 234–251.

Rosen, S. (1981). The Economics of Superstars. *The American Economic Review, 71*(5), 845–858.

Stiner, M. C. (2002). Carnivory, Coevolution, and the Geographic Spread of the Genus Homo. *Journal of Archaeological Research, 10*(1), 1–63.

Tarnita, C., Antal, T., Ohtsuki, H., Nowak, M., & May, R. (2009). Evolutionary Dynamics in Set Structured Populations. *Proceedings of the National Academy of Sciences of the United States of America, 106*(21), 8601–8604.

Thurber, B. A. (2015). The Viking Ball Game. *Scandinavian Studies, 87*(2), 167–188.

von Rueden, C., Gurven, M., & Kaplan, H. (2011). Why Do Men Seek Status? Fitness Payoffs to Dominance and Prestige. *Proceedings: Biological Sciences, 278*(1715): 2223–2232.

Wilson, J. (1931). Ceremonial Games of the New Kingdom. *The Journal of Egyptian Archaeology, 17*(3/4), 211–220.

Epilogue

Looking (at) Ahead

It is, perhaps, worthwhile stepping back and reflecting on the broader picture suggested by the theory of ideas presented in this book. Taken as a whole, this broad picture is infinitely more complex than perhaps even the tree of life. Current best estimates place the number of eukaryotic species on the planet to be roughly 9 million (Mora et al. 2011). Any best estimate for the number of ideas, by contrast, would be a meaningless concept. The 'tree of ideas'—perhaps not a very useful image—does not have definite termini; however, like the tree of life, it does have several common branches.

It is these common branches that we have tried to conceptualize in the theory of ideas and subsequently explore in our search for the collective idea space for sports. We have argued that sitting atop the hierarchy of ideas are the core principles. These core principles are the basis for a foundational idea space, the relevance of which cannot be overstated. It is this foundational idea space that frames all other collective idea spaces that emerge at lower levels of aggregation than it. It defines cooperation and it contextualizes competition between ideas, both

P. Goorha and J. Potts, *Creativity and Innovation*,
https://doi.org/10.1007/978-3-319-94884-3

within and across diverse social groups. It moors collective idea spaces but permits them wide latitude, from their fuzzy ideas to their core mechanisms. The tree of life has the wonderful potential to make us feel connected with all forms of life that exist and have ever existed on our planet. In like manner, so too does the idea of a hierarchy of fuzzy rules—that begins for everyone with the foundational core principles—have the potential for us to identify fundamental common ground and, through the exploration of the features of our idea spaces, examine the sources for fundamental differences.

The broad explorations of the last two chapters on the core ideas for some sports have served to bring light to the larger fact that core ideas for broad collective idea spaces are, in a manner of speaking, earned. The brief stories we considered on the development of lithic technology, hunting and herding all suggest that the principal benefit that evolution conferred upon humans was that the first and most enduring collective idea spaces developed organically on the basis of core principles being replicated across individual idea spaces, and only when a mechanism emerged that was closely allied to these commonly held crisp ideas.

Let us consider a counter example—artificially intelligent robots. Robots have been used for industrial manufacturing purposes for many decades. What has changed more recently, and indeed gathered pace very quickly, is the fact that artificial intelligence algorithms have endowed mechanical robots with the capability for learning through their interactions.

We now have robots that can mimic all the complexities of the human body. They can throw objects exceedingly accurately and be taught to catch them; there are those that can repeatedly strike with accuracy; and there are robots that can be collected into mobile or stationary teams and designed to learn tasks in such a manner that they then self-organize and specialize into subtasks. The Internet is abuzz with scores of articles, images, and entertaining videos of robots executing incredible feats of dexterity.

It is hard not to be impressed by Atlas, for example, which was developed by Boston Dynamics and the US Defense Advanced Research Projects Agency. Since its debut in 2013, Atlas has become increasing capable at walking in complex environments, interacting with a range of objects, jumping, pivoting midair, and even executing backflips. Watching ASIMO, the robot developed by Honda, do sign language, open a bottle, play soccer, and interact with humans in busy cities is perhaps even more astounding. And, in the context of our particular application to sports in this book, it is hard not to be amazed by Forpheus, a robot developed by Omron, a Japanese robotics company, that plays table tennis by learning the capabilities of its different human opponents by evaluating them through a range of cameras and, thereafter, engaging them in a rally of an appropriate difficulty.

While these advances, and a host of others, are indeed astounding as technological achievements, they have made several people very uncomfortable about the prospect of artificially intelligent machines becoming ubiquitous among us, as well as being directly integrated into the artificially intelligent software that is already part of our lives, such as in our phones, TVs, cars, social media platforms, and much else.

What is the source of this unease?

There is a range of objections that mostly hinge on the confidence that humans can retain control over these creations; or, as many fear, artificial intelligence will learn at rates unimagined by humans and beyond the capacities of our control.

More narrowly, in the context of robots, this unease can be examined in the context what has been called the 'uncanny valley'. Attributable to the robotics professor, Masahiro Mori, it is the observation that humans are favorable toward robots that do not resemble humans; develop a growing repulsion to robots that have an uncanny resemblance to humans in their facial features; and then finally emerge from this valley of this repulsion once the robot's appearance becomes indistinguishable from humans. There are several reasons for why this might be the case, but, from the perspective of our theory, a simple reason has less to do with the appearance of the robot, and more to do with the ideas that the robot is perceived to be capable of. When the robot's resemblance is almost like that of a human, the expectations are that the robot would have ideas that only mimic those of humans. Such mimicry might, of course, be a Trojan horse.

In other words, humans are instinctually on guard because they realize that the products of almost-human artificial intelligence may exhibit ideas that still lack a basis in the foundational idea space they subscribe to; in other words, such human-like ideas have not been *earned* through subsequent clarifications of core principles for particular contexts. It is perhaps why the test that psychologist, Robert Hare, developed for diagnosing human psychopaths emphasizes how psychopaths only mimic the appearance of normality in their behaviors; or, in other words, they, too, are almost human.

The general point is that while, locally, instances of several artificially intelligent robots are arguably bound by the foundational idea spaces of their creators, this certainly need not be the case globally, which is to say across all artificial intelligence. A recent study makes this point in a very vivid way in the context of video games. In a reward-poor game environment, namely a game that gives few cues on the quality of your immediate progress, it showed that humans tend to be considerably better than the sort of learning algorithms that drive artificial intelligence because they are able to rely on prior information (Dubey et al. 2018). What is this prior information in the context of artificial intelligence is less well-defined; indeed, it is exceedingly poorly defined if one were to attempt to correlate those ideas with those held by all humans.

The way forward, we suggest, is to ensure that the core principles of the foundational idea space for humans are reified across all applications of artificial intelligence. Our own diversity of ideas is proof that this does not imply a stifling of the idea spaces that can result from this promising new technology. Moreover, by having the foundational idea space be common to humans and artificial intelligence, we stand a better chance at being able to integrate more effectively with the products, both imagined and as yet unimagined, that this technology is likely to yield.

References

Dubey, R., Agrawal, P., Pathak, D., Griffths, T. L., & Efros, A. A. (2018). Investigating Human Priors for Playing Video Games. *arXiv*:1802.10217.

Mora, C., Tittensor, D. P., Adl, S., Simpson, A. G. B., & Worm, B. (2011). How Many Species Are There on Earth and in the Ocean? *PLoS Biology, 9*(8), e1001127.

Index

P. Goorha and J. Potts, *Creativity and Innovation*,
https://doi.org/10.1007/978-3-319-94884-3

The manufacturer's authorised representative in the EU is Springer Nature Customer Service Centre GmbH, Europaplatz 3, 69115 Heidelberg, Germany. If you have any concerns regarding our products, please contact ProductSafety@springernature.com

Printed and bound by CPI Group (UK) Ltd, Croydon, CR0 4YY
22/09/2026
02220950-0001